SpringerBriefs in Electrical and Computer Engineering

Cooperating Objects

Series Editor

Pedro José Marron

For further volumes:
http://www.springer.com/series/10208

Nouha Baccour · Anis Koubâa
Carlo Alberto Boano · Luca Mottola
Hossein Fotouhi · Mário Alves
Habib Youssef · Marco Antonio Zúñiga
Daniele Puccinelli · Thiemo Voigt
Kay Römer · Claro Noda

Radio Link Quality Estimation in Low-Power Wireless Networks

Nouha Baccour
Hossein Fotouhi
Claro Noda
Instituto Superior de Engenharia do Porto
Porto
Portugal

Anis Koubâa
Mário Alves
Instituto Superior de CISTER Research Centre
Porto
Portugal

Carlo Alberto Boano
Kay Römer
Universtät Lübeck
Lübeck
Germany

Luca Mottola
Dipartimento di Elettronica ed
Informazione, Politecnico di Milano
Milan
Italy

Habib Youssef
University de Sousse
Sousse
Tunisia

Marco Antonio Zúñiga
Universität Duisburg-Essen
Essen
Germany

Daniele Puccinelli
University of Applied Sciences and Arts
of Southern Switzerland
Manno
Switzerland

Thiemo Voigt
Swedish Institute of Computer Science AB
Kista
Sweden

ISSN 2191-8112 ISSN 2191-8120 (electronic)
ISBN 978-3-319-00773-1 ISBN 978-3-319-00774-8 (eBook)
DOI 10.1007/978-3-319-00774-8
Springer Cham Heidelberg New York Dordrecht London

Library of Congress Control Number: 2013940082

Printed on acid-free paper

Springer is part of Springer Science+Business Media (www.springer.com)

Preface

Stringent cost and energy constraints in low-power wireless networks (such as wireless sensor networks) impose the use of low-cost radio transceivers that transmit low-power signals (typically, 0 dBm as maximum power). This fact limits the radio channel, making it more vulnerable to noise, interference, and multipath distortion.

Moreover, these radio transceivers typically rely on inexpensive and size constrained antennas. Often, infeasibility of prime antenna positioning further compromises its performance. This applies to both, antenna positioning in certain WSN node designs as well as node placement in deployments, leading to anisotropic connectivity.

Consequently, low-power radio links are extremely unreliable and often unpredictable. They experience quality fluctuation over time and space, and show asymmetric connectivity. The unreliability of links greatly affects the network performance. This raised the need for link quality estimation as a fundamental building block for network protocols and mechanisms (e.g., medium-access control (MAC), routing, mobility management, and topology control), in order to mitigate link unreliability.

Link quality estimation in low-power wireless networks is a challenging research problem due to the lossy and dynamic nature of the links. This book aims at providing a good understanding of several aspects of link quality estimations, which covers the design, evaluation, experimentation, and impact on higher layer protocols.

Organization of the Book

This book is organized into five chapters. Chapter 1 outlines the characteristics of low-power links. A vast array of research efforts tackled the empirical characterization of low-power links through real-world measurements with different platforms, under varying experimental conditions, assumptions, and scenarios. This first chapter provides a comprehensive survey of the most relevant key observations drawn from empirical studies on low-power links. Chapter 2 addresses the most important factors that affect low-power links, which is interference. This chapter discusses several aspects related to interference problems in low-power wireless networks, namely experimentation, measurement, modeling, and mitigation of external radio interference. Chapter 3 gives an overview of link quality estimation in low-power wireless networks. It starts by presenting the fundamental concepts related to link quality estimation. For example, it defines the link quality estimation process and decomposes it into different steps. It also provides important requirements for the design of efficient link quality estimators (LQEs). Then, a taxonomy of existing LQEs is given. Chapter 4 provides an extensive comparative performance study of most well-known LQEs, based on both simulation and real experimentation. The evaluation methodology consists in analysing the statistical properties of LQEs, independently of any external factor, such as collisions and routing. These statistical properties impact the performance of LQEs, in terms of reliability and stability. Chapter 5 investigates the impact of using reliable and efficient LQEs for improving higher layer protocols and mechanisms, namely routing and mobility management.

This book is the result of several years of research work of the authors on the link quality estimation and low-power radio links. To the best of our knowledge, this is the first book tackling link quality estimation in low-power wireless networks (such as wireless sensor/actuator networks). Thus, we hope it will serve as baseline reading material for students and as a reference text to orient researchers and system designers.

Contents

Acronyms

AC	Supply line
ACK	Acknowledgement
AFH	Adaptive Frequency Hopping
ALQI	Average LQI
AP	Access Point
ARQ	Automatic Repeat reQuest
ARR	Acknowledgment Reception Ratio
ASK	Amplitude Shift Keying
ASL	Link Asymmetry Level
ASNR	Average Signal to Noise Ratio
AWGN	Additive White Gaussian Noise
BPM	Burst Position Modulation
BPSK	Binary Phase-Shift Keying
CCA	Clear Channel Assessment
CDF	Cumulative Distribution Function
CPDF	Conditional Probability Distribution Function
CPM	Closest-fit Pattern Matching
CRC	Cyclic Redundancy Check
CSI	Channel State Information
CSMA-CA	Carrier Sense Multiple Access with Collision Avoidance
CSS	Chirp Spread Spectrum
CTP	Collection Tree Protocol
CV	Coefficient of Variation
DataAnlApp	Data Analysis Matlab application
DCF	Distributed Coordination Function
DIFS	Distributed Inter-Frame Space
DQCSK	Differential Quadrature Chirp-Shift Keying
DSSS	Direct Sequence Spread Spectrum
DUCHY	DoUble Cost Field HYbrid
ETX	Expected Transmission Count
EWMA	Exponentially Weighted Moving Average
ExpCtrApp	Experiment Control java application
FEC	Forward Error Correction

FHSS	Frequency Hopping Spread Spectrum
FLQE-RM	Fuzzy Link Quality Estimator based Routing Metric
GFSK	Gaussian Frequency-Shift Keying
GSM	Global System for Mobile communications
HM	Hysteresis Margin
ICMP	Internet Control Message Protocol
IP	Internet Protocol
IPI	Inter-Packet-Interval
ISM	Industrial, Scientific, and Medical
KLE	Kalman filter-based Link quality Estimator
LEEP	Link Estimation Exchange Protocol
LETX	LQI-based ETX
LI	Link Inefficiency
LQE	Link Quality Estimator
LQI	Link Quality Indicator
MAC	Medium Access Control
MN	Mobile Node
MPSK	M-ary Phase Shift Keying
NACK	Negative-Acknowledgment
O-QPSK	Offset Quadrature Phase-Shift Keying
ParentCh	Average number of Parent Changes per node
PCB	Printed Circuit Board
PDR	Packet Delivery Ratio
PER	Packet Error Rate
PHY	Physical Layer
PRR	Packet Reception Rate
PSR	Packet Success
PSSS	Parallel Sequence Spread Spectrum
RA	Router Advertisement
RF	Radio Frequency
RFID	Radio Frequency IDentification
RNP	Required Number of Packet retransmissions
RS	Reed-Solomon
RSS	Received Signal Strength
RSSI	Received Signal Strength Indicator
RTX	Average number of setransmissions across the network per delivered packet
SDR	Software Defined Radio
SF	Spectrum Factor
SF	Stability Factor
SFD	Start-of-Frame Delimiter
SH-WSN6	Soft Handoff 6lowpan
SIFS	Short Inter-Frame Space
SINR	Signal-to-Interference-plus-Noise-Ratio
SNR	Signal-to-Noise Ratio

SOC	System On Chip
SPRR	Smoothed Packet Reception Ratio
TCP	Transmission Control Protocol
TH	Threshold Level
TSMP	Time Synchronized Mesh Protocol
UDP	User Datagram Protocol
UI	User-Interface
URN	Uniform Random Number
USRP	Universal Software Radio Peripheral
UWB	Ultra-Wide Band
WMEWMA	Window Mean with Exponentially Weighted Moving Average
WRE	Weighted Regression Estimator
WSN	Wireless Sensor Network

Chapter 1
Characteristics of Low-Power Links

Abstract This chapter aims at providing a summary of key properties of low-power links. First, it gives an overview of the radio technology most commonly used in low-power wireless networks. Then, it distils from the vast array of empirical studies on low-power links a set of *high-level observations*, which are classified into spatial and temporal characteristics, link asymmetry, and interference. Such observations are helpful not only to design efficient Link Quality Estimators (LQEs) that take into account the most important aspects affecting link quality, but also to design efficient network protocols that have to handle link unreliability.

1.1 Introduction

For an ideal non-obstructive environment, the nature of electromagnetic waves propagation attenuates the signal power abruptly near the transmitter and yields much less attenuation at longer distances. This is described by the Friis Equation, directly derived from fundamental theory.

In a real environment, obstacles in the signal path absorb energy and the degree of signal attenuation with distance further increases. Thus, the signal pathloss can be phenomenologically described by the log-normal shadowing propagation model. Depending on the specific environment, the pathloss exponent would vary between 2 and 6.

As a consequence, a low-power radio link operates on an RF power budget which drops rapidly with the distance between sender and receiver nodes. The lower link margin provides a lower Signal to Noise Ratio (SNR) which (theoretically) determines the Packet Reception Rate (PRR), producing an intrinsic less stable (lower quality) short-range link, compared to the links in other wireless systems.

The propagation of radio signals is also affected by other factors that contribute to link quality degradation. These factors include (i) the multi-path propagation effect, which depends on the environment, (ii) the interference, which results from

N. Baccour et al., *Radio Link Quality Estimation in Low-Power Wireless Networks*,
SpringerBriefs in Electrical and Computer Engineering,
DOI: 10.1007/978-3-319-00774-8_1,

concurrent transmissions within the wireless network or between cohabiting wireless networks and other electromagnetic sources; and (iii) hardware transceivers, which may distort transmitted and received signals due to their internal noise.

Several research efforts have been devoted to an empirical characterization of low-power links. These studies have been carried out using (i) different platforms having different radio chips (TR1000, CC1000, CC2420, etc), (ii) different operational environments (indoor, outdoor) and (iii) different experimental settings (e.g., traffic load, channel). Therefore, they presented radically different (and sometimes contradicting) results. Nonetheless, these studies commonly argued that low-power links exhibit complex and dynamic behaviours.

Although several low-power link characteristics are shared with those of traditional wireless networks such as ad-hoc, mesh, and cellular networks, the extent of these characteristics is more significant with low-power links (e.g, a large transitional region or extremely dynamic links) and makes them even more unreliable. This might be an artifact of the communication hardware used in low-power wireless networks [1, 2].

Next, we synthesize the vast array of empirical studies on low-power links and draw a set of *high-level observations*. Beforehand, we present an overview of the most common radio transceivers in low-power wireless networks as they represent a major cause of low-power links unreliability.

1.2 Low-Power Wireless Hardware

The hardware enabling wireless transmissions bears a direct impact on the characteristics of wireless links. Two key components concur to these factors: the specific radio and the antenna technology. In the following, we briefly survey the most common hardware platforms available, laying out the technological basis for the material ahead.

1.2.1 Radios

The specific radio chip employed is a key factor in determining the characteristics of wireless links. Table 1.1 illustrates the characteristics of radio platforms most commonly used in low-power wireless networks, along with a few examples of next-generation radio platforms that may gain significant adoption.

To tackle the energy issue, early hardware platforms such as ChipCon CC1000 and RFM TR1000 leveraged radio chips operating in sub-GHz frequencies. These transceivers offer low power consumption in both transmission and receive modes. On the other hand, the low data rate prevented using these devices in scenarios different from low-rate data collection.

The need for higher data rate motivated the design of radios working in the 2.4 GHz ISM band, such as the widely used ChipCon's CC2400 and CC2500 families.

Table 1.1 Characteristics of sample WSN radios

Model	Frequency	Max data rate	Modulation	TX current	RX current	Max TX power
CC1000	300–1000 MHz	76.8 kbps	2-FSK	18.5 mA	9.6 mA	10 dBm
nRF903	433 or 915 Mhz	76.8 kbps	GFSK	19.5 mA	22.5 mA	10 dBm
TR1000	916 Mhz	115.1 kbps	OOK/ASK	12 mA	3.8 mA	0 dBm
CC2420	2.4 Ghz	250 kbps	DSSS/O-QPSK	17.4 mA	19.7 mA	0 dBm
CC2500	2.4 Ghz	512 kbps	2-FSK	12.8 mA	21.6 mA	1 dBm
PH2401	2.4 Ghz	1 mbps	GFSK	<20 mA	<20 mA	2 dBm
EFR4D2290 (SOC platform)	2.4 GHz	4 mbps	flexible	6 mA	4.5 mA	< +13 dBm
GS1011 (Low-power Wifi)	2.4 GHz	1–11 mbps	DSS/BPSK/CCK	140 mA	150 mA	< +18 dBm

Compliance to IEEE 802.15.4 also fostered a wider adoption of these radio chips, which are commonly found in several current WSN platforms.

The tendency to provide higher data rates is brought to an extreme when Bluetooth or WiFi chips are used. These may be found in hybrid configurations where a high-data rate radio is coupled to a low-power one. For instance, the BTnode [3] platform uses a Bluetooth-compliant device next to a CC1000 chip. Such design allows greater flexibility and alternative uses of the WSN devices, e.g., as passive sniffers of ongoing traffic for debugging purposes [4].

Nevertheless, recent advancements in low-power chip design yield configurations where WiFi chips are deployed in stand-alone modules [5] or a System-on-Chip (SoC) [6], achieving reasonable lifetimes in specific scenarios. As a result, such radios are particularly amenable, for example, for bursty point-to-point traffic.

Low power WiFi radios, offer lower Energy-per-useful-bit and higher datarates, compared to 802.15.4 radios. More resourceful radios, however, are not suitable for all systems as the overall energy efficiency achievable also depends on node synchronization, duty-cycling and the application datarate requirements. For example, consider a network based on an asynchronous duty-cycled medium access control, which is common on sensor networks. The energy overhead as nodes rendezvous, would outweigh the benefits of a more resourceful radio, which consumes more power on active mode.

Additionally, an SoC features high level of integration and tight interconnection among all subsystems, which contribute to increased performance and energy-efficiency. In a SoC the radio and the Micro Controller Unit (MCU) share buffered data registers in RAM and a DMA controller allows for data transfers without software intervention, even with the main MCU in a deep sleep mode.

The Energy Friendly Radios (EFR) series from Energy Micro^{TM}, are based on a 32-bit ARM Cortex RISC processor [7]. These SoC solutions include radios that operate in a wide range of bands and represent the most energy-efficient radios to reach the market [8], at the time of writing.

1.2.2 Antennas

Complementing the radio hardware is the specific antenna enabling the actual transmissions over the air. The coupling of the two determines to a large extend the characteristics of the resulting wireless links.

Sensor devices are often shipped with low-gain antennas integrated in the board. For instance, in the widespread TMote/TelosB devices (Fig. 1.1a) [9], the antenna is integrated in the PCB (Printed Circuit Board). In such a design, although intended to be omni-directional, the actual radiation pattern results irregular (Fig. 1.1b) mainly because of the presence of the node circuitry and battery package close to the antenna. These aspects complicate the operation of MAC and routing protocols, which are traditionally based on the assumption of uniform communication ranges and symmetric links.

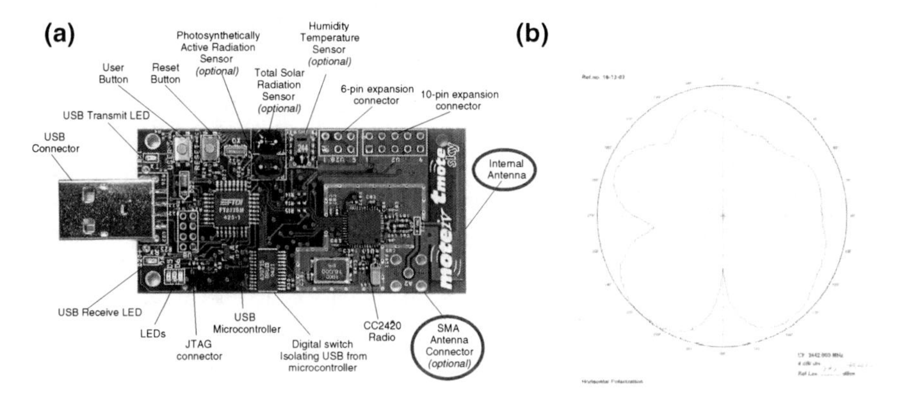

Fig. 1.1 TMote antenna details. **a** Integrated micro-strip antenna. **b** Radiation pattern with horizontal mounting

To remedy these issues and for better performance, a common design choice in real-world deployments is the replacement of the standard antenna [10] with custom one. This brings increased communication range and higher reliability without incurring extra energy costs. For instance, antennas of up to 8.5 dBi were used in harsh environments by exploiting the on-board SMA connectors [11].

Fixed-direction antennas, which are able to direct the transmission power in directions determined at construction time, have also been used in real deployments and as debugging tools. For instance, Saukh et al. [12] use "cantennas"—simple cylinder-shaped directional antennas—for node localisation and selective communication to a group of nodes. However, for normal operation they typically lack flexibility in freely reconfiguring the network topology and node locations [13].

A new breed of controllable directional antennas are also emerging. Nilsson identifies three such classes for low-power networks [14]: the adcock-pair antenna, the pseudo-doppler antenna, and the electronically switched parasitic element antenna. For example, Giorgietti et al. [15] describe a prototype of four-beam patch antenna, whereas Nilsson also designs the SPIDA antenna, an electronically switched parasitic element antenna. Both antennas are integrated with the CC2420 radio chip aboard TMote Sky nodes [16] and the direction of maximum gain is software-controlled. Support of networking protocols able to take advantage of such features is, however, largely missing at present.

1.3 Link Quality Metrics

In this section, we briefly present a set of basic metrics that were examined by empirical studies to capture low-power link characteristics:

- **PRR** (Packet Reception Ratio)—sometimes referred to as PSR (Packet Success Ratio). It is computed as the ratio of the number of successfully received packets to the number of transmitted packets. A similar metric to the PRR is the PER (Packet Error Ratio), which is *1 - PRR*.
- **RSSI** (Received Signal Strength Indicator). Most radio transceivers (e.g., the CC2420) provide a *RSSI register*. This register provides the signal strength of the received packet. When there are no transmissions, the register gives the noise floor.
- **SNR** (Signal to Noise Ratio). It is typically given by the difference in decibel between the pure (i.e., without noise) received signal strength and the noise floor.
- **LQI** (Link Quality Indicator). It is proposed in the IEEE 802.15 standard [19], but its implementation is vendor-specific. For the CC2420 [20], which is the most widespread radio, LQI is measured based on the first eight symbols of the received packet as a score ranging from 50 to 110 (higher values are better).

Next, we analyse the empirical characterization of low-power links, which can be classified into three parts: Spatial characteristics, temporal characteristics, link asymmetry, and interference.

1.4 Spatial Characteristics

It was demonstrated in several studies that the transmission range is not isotropic (i.e., a circular shape), where packets are received only within a certain distance from the sender [21]. In fact, the transmission range is defined by three regions; each with an irregular shape, dynamic bounds (changing over the time), and specific features [18, 22–24]. These regions are: (i) connected region, where links are often of good quality, stable, and symmetric, (ii) transitional region, where links are of intermediate quality (in long-term assessment), unstable, not correlated with distance, and often asymmetric, and (iii) disconnected region, where links have poor quality and are inadequate for communication. Particularly, the transitional region was subject of several empirical studies because links within this region are extremely unreliable and even unpredictable [1, 18, 22–24]. These intermediate quality links, referred also as *intermediate links*, are commonly defined as links having an average PRR between 10 and 90 %.

Observation 1 Link quality is not correlated with distance, especially in the transitional region. To observe the transitional region, most empirical studies conducted measurements of the PRR at different distances from the sender. Figure 1.2b is an illustration of the three communication regions through PRR measurements. This figure shows that link quality is *not correlated with distance*, especially in the transitional region. Indeed, two receivers placed at the same distance from the sender can have different PRRs, and a receiver that is farther from the sender can have higher PRR than another receiver nearer to the sender. This observation can be clearly understood from Fig. 1.2a.

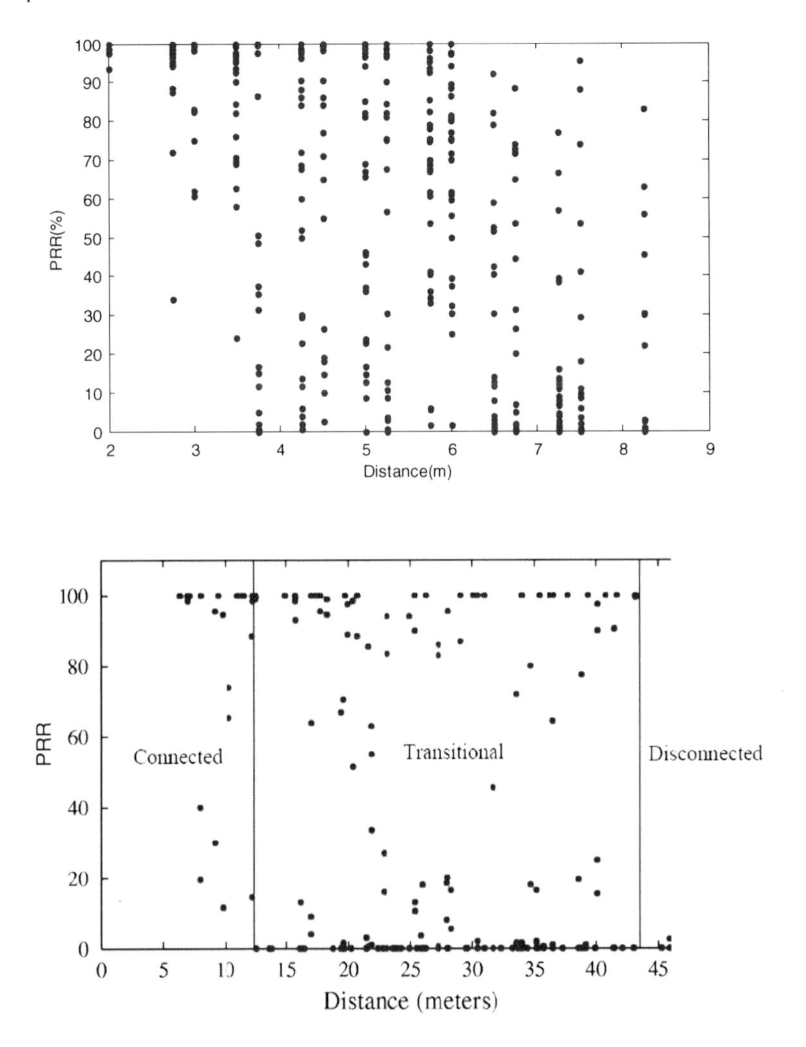

Fig. 1.2 Spatial characteristics: PRR as a function of distance between receiver node and sender node. **a** The transitional region, for an outdoor environment, using TelosB sensor motes and −25 dBm as output power (using the RadiaLE testbed [17]). **b** The *three* reception regions, for an outdoor Habitat environment, using Mica 2 sensor motes and −10 dBm as output power [18]

Observation 2 The extent of the transitional region depends on (i) the environment (e.g., outdoor, indoor, presence of obstacles), and (ii) the radio hardware characteristics (e.g., the transmission power, the modulation schema, the radio chip) [25]. However, the quantification of this extent by empirical studies shows contradicting observations. Measurements of PRR according to distance, for different environments, radios, and power settings were carried out. In [18], Cerpa et al. performed measurements in indoor (Office) and outdoor (Habitat) environments using the Mica

1 and Mica 2 platforms and different power levels, namely −10 dBm, −6 dBm and 1 dBm. They found that the width of the transitional region is significant, ranging from 50 % up to 80 % of the transmission range. On the other hand, Zhao et al. [22] performed measurements with almost the same settings as of Cerpa et al. [18], but they found the transitional region width smaller, almost one-fifth up to one-third of the transmission range.

Observation 3 The percentage of intermediate quality links (i.e., located in the transitional region) was found significant in some empirical studies and insignificant in others, which lead to contradicting results. In [22], Zhao et al. performed experiments with Mica 1 platform in an office building while varying the traffic load. They found that the percentage of intermediate quality links ranges from 35 to 50 %. On the other hand, Srinivasan et al. [1] performed experiments with more recent platforms, Micaz and TelosB, in different environments and with varying traffic loads. They found that the number of intermediate links ranges from 5 to 60 %. Based on this observation, they claimed that the number of intermediate links observed with recent platforms is lower than that observed with old platforms. This was justified by the fact that recent platforms integrate IEEE 802.15.4 compliant radios (e.g., the CC2420) that have more advanced modulation schemes (e.g., Direct Sequence Spread Spectrum (DSSS)) compared to old platforms. Mottola et al. [26] refuted this observation while conducting experiments in road tunnels using motes having IEEE 802.15.4 compliant radios. They observed a large transitional area in two of their tunnels and found a high number of intermediate quality links due to the constructive/destructive interference. We believe that this aspect remains an open issue and needs to be supported by additional experiments for two reasons. First, intermediate quality links were defined differently, namely "links with PRR less than 50 %" in [22] and "links with PRR between 10 and 90 %" in [1]. Second, experimental studies that analyzed the percentage of intermediate quality links were based on different network settings (e.g., traffic load, power level, radio type, and environment type) and also different window sizes for PRR calculation, so a comparison is not completely legitimate.

Observation 4 Link quality is anisotropic. Empirical studies observed another important spatial characteristic of low-power links often referred as *radio irregularity*, which means that link quality is anisotropic [23, 27–30]. To demonstrate the existence of radio irregularities, Zhou et al. [30] observed the RSSI and the PRR according to different receiver's directions, but with fixed distance between the transmitter and the receiver. They showed that the radio communication range, assessed by the RSSI, exhibits a non-spherical pattern. They also argued the existence of a non-spherical interference range, located beyond the communication range (refer to Fig. 1.3). Within this interference range the receiver cannot interpret correctly the received signal, but this received signal can prevent it from communicating with other transmitters as it causes interference. The existence of the non-spherical radio communication and interference ranges was confirmed by Zhou et al. [29]. They reported that in WSNs, several MAC protocols assume the following: If node B's signal can interfere with node A's signal, preventing A's signal from being received at node C; then node C must be within node B's communication range. Based on experimentation with Mica 2 motes, Zhou et al. showed that this assumption is definitely invalid,

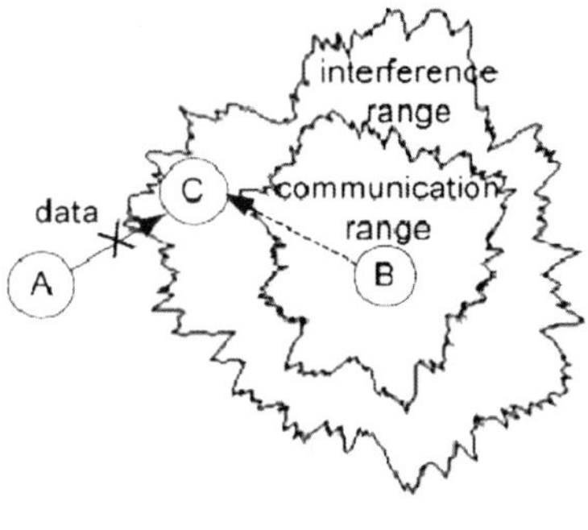

Fig. 1.3 Radio irregularity and interference range [27]. *Node B* cannot communicate with *node C* as it is out of its communication range. However, *B* prevents *C* to communicate with *A* due to the interference between the signal sent by *B* and that sent by *A*

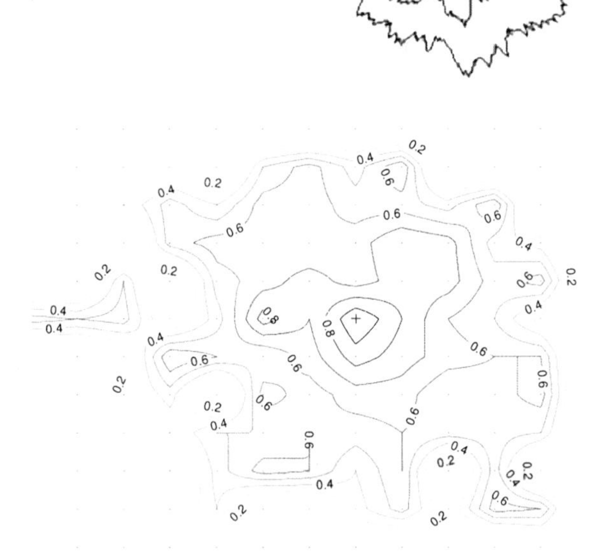

Fig. 1.4 Contour of PRR from a central node: anisotropy of link quality [28]

since node C may be in the interference range of node B and not in its communication range, as illustrated in Fig. 1.3. The communication range assessed by the PRR was also shown to be non-spherical or anisotropic [28], as shown in Fig. 1.4. A natural reason for radio irregularity is the anisotropic radiation pattern of the antenna due to the fact that antennas do not have the same gain along all propagation directions [30].

Observation 5 Sensor nodes that are geographically close to each other may have high spatial correlation in PRRs. Zhao et al. [22] investigated the spatial correlation in PRRs, measured between a source node and different receiver nodes. They observed that receiver nodes that are geographically close to each other and that are located in the transitional region, have higher coefficient of correlation in their PRRs, compared to nearby receiver nodes located in the connected or disconnected regions. Nevertheless, the coefficient of correlation in the transitional region is not that significant—less than 0.7. Srinivasan et al. [31] introduced the κ Factor, a new metric that captures spatial correlation in PRR between links, using the cross-correlation index. The κ Factor was shown to perform better than exiting metrics for the measurement of spatial correlation between links.

Observation 6 The spatial variation of link quality is due to constructive/destructive interference. Beyond the connected region, the direct signal is weak due to path loss. Multi-path effects can be either *constructive*, i.e., strengthen the direct signal leading to a good quality link, or *destructive*, i.e., interfere with the direct signal [18], and thus be detrimental to link quality. Being constructive or destructive does not depend on the receiver distance or direction. It rather depends on the nature of the physical path between the sender and the receiver (e.g., presence of obstructions) [32, 33].

1.5 Temporal Characteristics

We showed that link quality varies drastically over space. This section explores link quality variation over time.

Observation 1 Links with very low or very high average PRRs are more stable than links with moderate average PRRs. Several studies [18, 22, 34] claimed that links with very low or very high average PRR, which are mainly located in the connected and disconnected regions respectively, have small variability over time and tend to be stable. In contrast, links with intermediate values of average PRR, which are mainly located in the transitional region, show a very high variability over time, as PRRs vary drastically from 0 to 100 % with an average ranging from 20 to 80 % [18]. These intermediate links are hence typically unstable. This observation is illustrated in Fig. 1.5. The temporal variation of these links can be mitigated by applying an adaptive power control scheme, where transmission power at each node is dynamically adjusted [35].

Observation 2 Over short time spans, links may experience high temporal correlation in packets reception, which leads to short periods of 0 % PRR or 100 % PRR. Srinivasan et al. [1] examined the distribution of PRRs over all links in the test-bed, for different Inter-Packet-Intervals (IPIs). They found that by increasing the IPI, the number of intermediate links increases as well. This finding was justified by the fact that low IPIs correspond to a short-term assessment of the link. In such short-term assessment, most links experience high temporal correlation in packets reception. That means that over these links, packets are either all received or not. Consequently, the measured PRR over most links is either 100 or 0 %. For instance, Srinivasan et al. [1] found that for a low IPI equal to 10 milliseconds (PRRs are measured every 2 s) 95 % of links have either perfect quality (100 % PRR) or poor quality (0 % PRR), i.e., only 5 % of links have intermediate quality. High IPIs corresponds to a long-term assessment of the link. The increase of the IPI leads to the decrease in the temporal correlation in packets reception. That means that links may experience *bursts* (a shift between 0 and 100 % PRR) over short periods and the resulting PRR assessed in long-term period is intermediate. This last observation was also made in [36].

Recently, several metrics were introduced for the measurement of link burstiness. Munir et al. [37] define a burst as a period of continuous packet loss. They introduced *Bmax*, a metric that computes the maximum burst length for a link, i.e., the maximum number of consecutive transmission failures. *Bmax* is computed using an algorithm

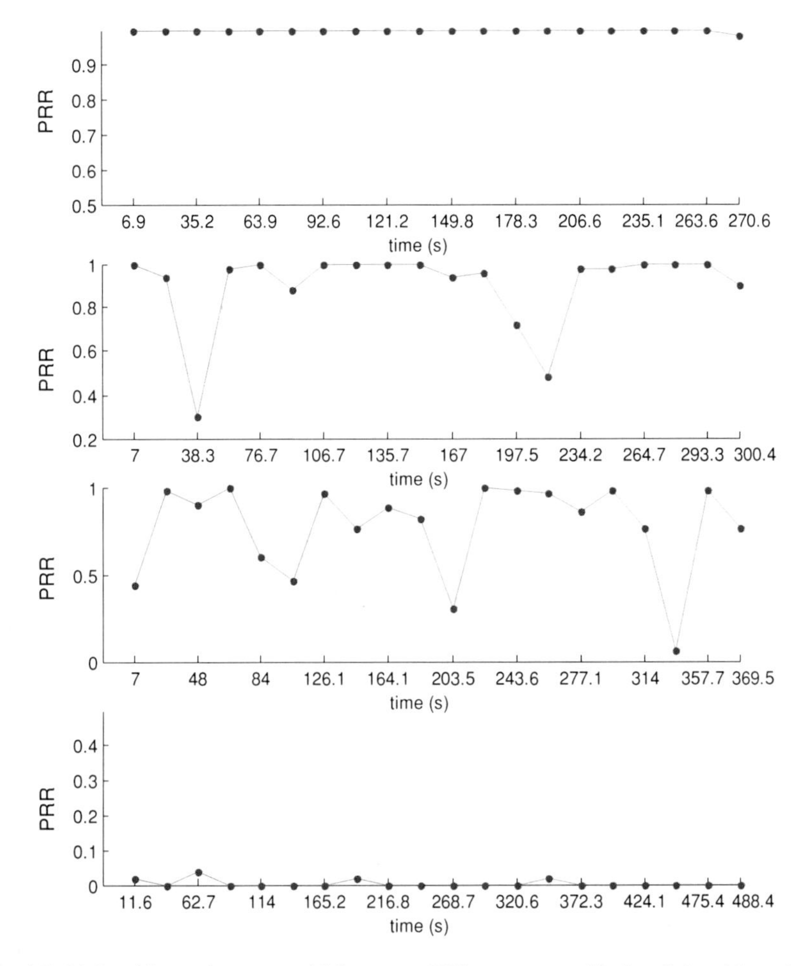

Fig. 1.5 Links with very low or very high average PRRs are more stable than links with moderate average PRRs. Outdoor environment, using TelosB sensor motes and −25 dBm as output power (using RadiaLE testbed [17])

that takes as input (i) the data trace of packet successes and failures for each link, and (ii) *B'min*, which is the minimum number of consecutive successful transmissions between two consecutive failure bursts. The authors assume a pre-deployment phase for the determination of *Bmax* with respect to each link in the network. However, computed *Bmax* values may change during the network operation due to environmental changes. In [38], the authors resolved this problem by introducing BurstProbe, a mechanism for assessing link burstiness through the computation of *Bmax* and *B'min* during the network operation. The β factor is another metric for assessing link burstiness [39]. It is used to identify bursty links with long bursts of successes or failures. The β factor is computed using conditional probability distribution functions (CPDFs), which determine the probability that the next packet will be received after *n* consecutive successes or failures. It requires a large data trace and thus might be inappropriate for online link burstiness assessment.

Observation 3 The temporal variation of link quality is due to changes in the environment characteristics. Several studies confirmed that the temporal variation of link quality is due to the changes in the environment characteristics, such as climate conditions (e.g., temperature, humidity), human presence, interference and obstacles [2, 22, 23, 36, 40, 41]. Particularly, in [2], the authors found that the temporal variation of LQI, RSSI, and Packet Error Rate (PER), in a "clean" environment, (i.e., indoor, with no moving obstacles and well air-conditioned) is not significant. The same observation was made in [26]. Lin et al. [41] distinguished three patterns for link quality temporal variation: small fluctuations, large fluctuations/disturbance, and continuous large fluctuations. The first is mainly caused by multi-path fading of wireless signals; the second is caused by shadowing effect of humans, doors, and other objects; and the last is caused by interference (e.g., Wi-Fi). A deeper analysis of the causes of links temporal variation was presented in [39, 42, 43]. Lal et al. [42] reported that the transitional region can be identified by the PRR/SNR curve using two thresholds (refer to Fig. 1.6). Above the first threshold the PRR is consistently high, about 100 %, and below the second threshold the PRR is often less than 25 %. In between is the transitional region, where a small variation in the SNR can cause a shift between good and bad quality link, which results in a bursty link. In fact, SNR is the ratio of the pure received signal strength to the noise floor. When no interference is present, the noise floor varies with temperature, and so is typically quite stable over time periods of seconds or even minutes [1]. Therefore, what makes the SNR vary according to time leading to link burstiness is mainly the received signal strength variation [39]. The variation of the received signal strength may also be due to the constructive/destructive interference in the deployment environment [26].

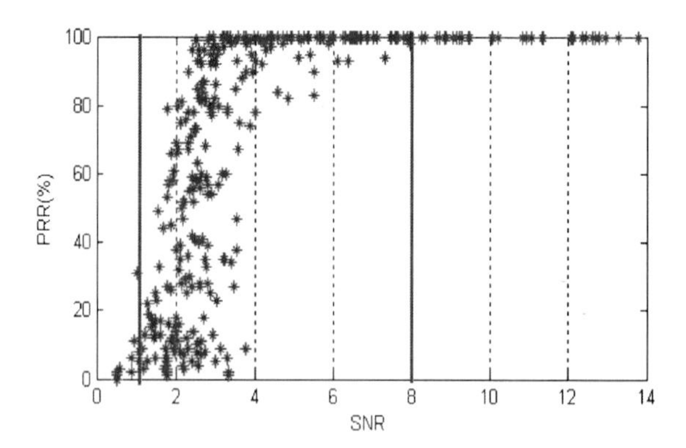

Fig. 1.6 The PRR/SNR curve. For *SNR* greater than 8 dBm, the *PRR* is equal to 100 %, and for *SNR* less than 1 dBm, the *PRR* is less than 25 %. In between, a small variation in the *SNR* can cause a big difference in the *PRR*; links are typically in the transitional region. Outdoor environment, using TelosB sensor motes and −25 dBm as output power (using the RadiaLE testbed [17])

1.6 Link Asymmetry

Link asymmetry is an important characteristic of radio links as it has a great impact on the performance of higher layer protocols. Several studies analyzed the asymmetry of low-power links [1, 2, 18, 22, 23, 25, 28, 34]. Link asymmetry is often assessed as the difference in connectivity between the uplink and the downlink. A wireless link is considered as asymmetric when this difference is larger than a certain threshold, e.g., when the difference between the uplink PRR and the downlink PRR is greater than 40 % [1, 18].

Observation 1 Asymmetric links are mainly located at the transitional region. It was shown that links with very high or very low average PRRs, which are mainly those of the connected and disconnected regions respectively, tend to be symmetric. On the other hand, links with moderate PRRs, those of the transitional regions, tend to be asymmetric [18, 34].

Observation 2 Link asymmetry is not correlated with distance. The spatial variation of link asymmetry was the subject of several studies [18, 23, 28, 34]. Ganesan et al. [28] found that the percentage of asymmetric links is negligible at short distances from the transmitter and increases significantly with higher distances. This observation confirms the one made by Cerpa et al. [18, 34], stating that asymmetric links are mainly those in the transitional region. On the other hand, Cerpa et al. found that the percentage of asymmetric links increases as well as decreases as the distance from the transmitter increases. Thus, they argued that link asymmetry is not correlated with distance.

Observation 3 Link asymmetry may or may not be persistent. Srinivasan et al. [1] studied the temporal variation of link asymmetry. They found that very few links (2 of the 16 observed asymmetric links in the testbed) were long-term asymmetric links (i.e., consistently asymmetric) while many links were transiently asymmetric. On the other hand, Mottola et al. [26] found that when links are stable, which is the case in their experiments, link asymmetry also tends to persist. Consequently, link asymmetry might be transient only for unstable links (i.e., their quality varies with time), and ultimately depends on the target environment.

Observation 4 Hardware asymmetry and radio irregularity constitute the major causes of link asymmetry. Most studies stated that one of the causes of link asymmetry is hardware asymmetry, i.e., the discrepancy in terms of hardware calibration; namely nodes do not have the same effective transmission power neither the same noise floor (receiver sensitivity) [18, 22, 25, 44]. Ganesan et al. [28] claimed that at large distances from the transmitter, small differences between nodes in hardware calibration may become significant, resulting in asymmetry. The radio irregularity caused by the fact that each antenna has its own radiation pattern that is not uniform, is another major cause of link asymmetry [30, 44].

1.7 Interference

Interference is a natural phenomena in wireless transmissions since the medium is shared among multiple transmitting nodes. In the following, we provide a bird's eye view on the current state-of-the-art related to interference in low-power wireless networks. Our goal is not to be exhaustive, but rather to give the reader a foundation to understand how interference may affect link quality estimation.

Interference can be either *external* or *internal*. External interference may occur from co-located/co-existing networks that operate in the same frequency band as the WSN; internal interference may occur from concurrent transmission of nodes belonging to the same WSN. In the rest of this section, we survey relevant work on both external and internal interference. Chapter 2 will be entirely dedicated to external interference, and to a detailed description of existing techniques for interference mitigation.

1.7.1 External Interference

WSNs operate on unlicensed ISM bands. Therefore they share the radio spectrum with several other devices. For example, in the 2.4 GHz frequency, WSNs might compete with the communications of Wi-Fi and Bluetooth devices. Furthermore, a set of domestic appliances such as cordless phones and microwave ovens generates electromagnetic noise which can significantly harm the quality of packet receptions [45–47]. External interference has a strong impact on the performance of WSN communications because it increases packet loss rate, which in turn increases the number of retransmissions and therefore the latency of communications.

Observation 1 The co-location of 802.15.4 and 802.11b networks affects transmission in both networks due to interference (unless the 802.15.4 network uses channel 26), but the transmission in 802.11b networks is less affected. Srinivasan et al. [1] observed that 802.11b transmissions (i.) can prevent clear channel assessment at 802.15.4 nodes, which increases latencies and (ii.) represent high power external noise sources for 802.15.4, which can lead to packet losses. They also observed that 802.11b nodes do not suspend transmission in the presence of 802.15.4 transmission, since 802.11b transmission power is 100 times larger than that of 802.15.4. However, this observation was refuted by Liang et al. [48]. Indeed, they reported that when the 802.15.4 transmitter is close to the 802.11b transmitter, the 802.11b node may suspend its transmission due to elevated channel energy. Furthermore, when this happens, IEEE 802.11b only corrupts the IEEE 802.15.4 packet header, i.e., the remainder of the packet is unaffected. The impact of interference generated by Wi-Fi devices strongly depends also on the traffic pattern. Boano et al. [49] presented experimental results using different Wi-Fi patterns and compared the different PRRs under interference. Srinivasan et al. [1] noticed that only 802.15.4 channel

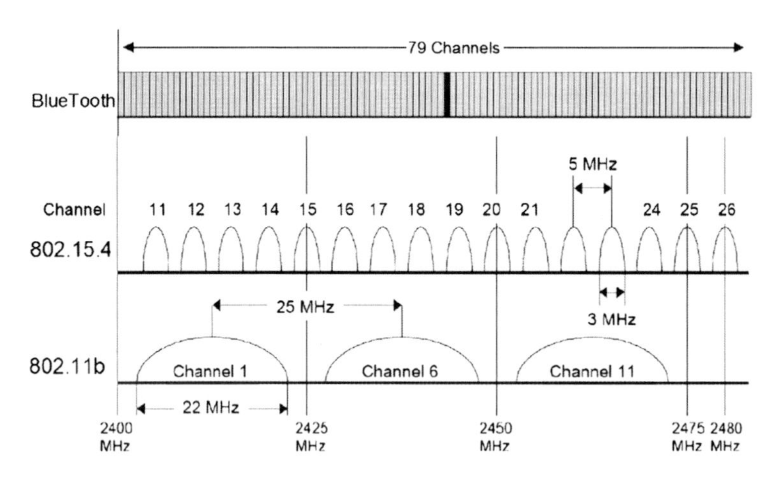

Fig. 1.7 IEEE 802.15.1 (Bluetooth), IEEE 802.11b, and IEEE 802.15.4 spectrum usage [1]

26 is largely immune to 802.11b interference, as it does not overlap with 802.11b channels (refer to Fig. 1.7).

Observation 2 The co-location of IEEE 802.15.4 and 802.15.1 (Bluetooth) networks affects mostly the transmissions in the IEEE 802.15.4 network. Bluetooth is based on frequency hopping spread spectrum (FHSS) technology. This technology consists in hopping to a new frequency after transmitting or receiving a packet, using a pseudorandom sequence of frequencies known to both transmitter and receiver. Thanks to this technology, Bluetooth is highly resistant to interference. Consequently, when 802.15.4 and Bluetooth networks coexist, packet losses at Bluetooth devices are not that important as compared to those observed with 802.15.4. The results by Boano et al. [49] show that interference from Bluetooth devices has a much lower impact than the one from Wi-Fi devices or microwave ovens on WSN communications.

Observation 3 The co-location of IEEE 802.15.4 networks and domestic appliances can significantly affect the transmission in the IEEE 802.15.4 networks. Using a spectrum analyzer, Zhou et al. [50] showed the impact of interference generated by a microwave oven, which can cover almost half of the 2.4 GHz available spectrum. Their results were confirmed by Boano et al. [49], who measured the periodic pattern of microwave ovens interference through fast RSSI sampling using off-the-shelf sensor motes. The authors highlighted the periodicity of the generated interference and quantified its impact on the PRR of WSN communications.

Observation 4 External interference often spreads along several (adjacent) channels [45, 46]. Due to the characteristics of external wide-band interferers such as Wi-Fi devices, interference often spreads throughout spatially nearby channels (refer to Fig. 1.7). Another example of the latter are microwave ovens, that spread noise over almost half of the 2.4 GHz available spectrum, as explained earlier.

1.7.2 Internal Interference

As external interference, internal interference can have a strong impact on the performance of WSN communications.

Observation 1 In the presence of concurrent transmission, the three reception regions can still be identified by the signal-to-interference-plus-noise-ratio (SINR). Most studies on low-power link characterization, including those stated previously, were performed using collisions-free scenarios to observe the pure behavior of the channel. In [51], the authors addressed low-power link characterization under concurrent transmissions. They reported that concurrent transmissions lead to interference, which has a great impact on link quality. Based on signal-to-interference-plus-noise-ratio (SINR) measurements, conducted with Mica 2 motes equipped with CC1000 radios, the authors found the following key observations: First, when the SINR exceeds a critical threshold, the link is of high quality,[1] i.e., the PRR is greater than 90 %, and it belongs to the connected region. Below this threshold, transmissions on that link can be successful despite the existence of concurrent transmissions, but the resulting PRR is below 90 % (transitional and disconnected regions). Second, the identified SINR threshold can vary significantly between different hardware. In fact, this threshold depends on the transmitter hardware and its transmission power level, but it does not depend on its location.

Observation 2 Concurrent transmissions have a great impact on the link delivery ratio even when nodes are not visible to each other. In [26], the authors conducted experiments in real road tunnels, with controlled concurrent transmissions. They set up a specific scenario where two nodes communicate and a third node, which is not visible to the first two (i.e., "far from" the two nodes and the PRR to each of them is equal to zero), concurrently transmits its data. They found that the third node was able to create a significant noise for the communicating nodes so that the delivery ratio over that link (assessed by the PRR) was very low, even lower than expected.

Observation 3 Internal Interference from adjacent channels has a significant influence on the packet delivery rate. Several work showed that cross-channel interference can cause a significant increase in the packet loss ratio [52–55]. Wu et al. [54] showed on MicaZ motes that with adjacent channel interference, the PRR decreases 40 % compared to when no interference is present on the adjacent channel. The authors also showed that when interference is generated two channels away, the impact on the PRR is minimal. Xing et al. [55] proposed an algorithm that reduces the overhead of multi-channel interference measurements by exploiting the spectral power density of the transmitter.

[1] This is interpreted by the fact that the strength of the received signal is much higher than those of the noise level and the received signal from the interfering node.

1.8 Conclusion

This chapter provides an overview of the most common radio technology in low-power wireless networks (such as wireless sensor networks) and the analysis of low-power link characteristics. We distilled from the vast array of empirical studies on low-power links a set of high-level observations, some of which are contradictory. This is mainly due to the discrepancies in experimental conditions of empirical studies, i.e., they do not have the same environment characteristics, the same platform, or the same experiment settings. Apart from these contradicting observations, we have identified a set of observations showing how low-power links experience a complex and dynamic behavior. In fact, low-power links are extremely unreliable due to the low-power and low-cost radio hardware typically employed in low-power nodes. Another important factor that contributes to link quality degradation is interference, which can be either external or internal. Due to its importance, the following chapter deals with external interference and outlines existing techniques for interference mitigation.

References

1. Srinivasan K, Dutta P, Tavakoli A, Levis P (2010) An empirical study of low-power wireless. ACM Trans Sen Netw 6:1–49
2. Tang L, Wang KC, Huang Y, Gu F (2007) Channel characterization and link quality assessment of ieee 802.15.4-compliant radio for factory environments. IEEE Trans Industr Inf 3(2):99–110
3. BTNode: http://www.btnode.ethz.ch
4. Dyer M, Beutel J, Thiele L, Kalt T, Oehen P, Martin K, Blum P (2007) Deployment support network—a toolkit for the development of WSNs. In: Proceedings of the 4th european conference on wireless sensor networks (EWSN '07). Springer, pp 195–211
5. Networks R (2012) RN-131 Datasheet. http://www.rovingnetworks.com/resources/download/11/RN_131
6. GainSpan: GS1011M Low-Power Wireless System-On-Chip WI-FI Module Data Sheet. http://www.gainspan.com/products/GS1011_SoC.php (2011)
7. Larsen RC, Janbu O (2009) Whitepaper: World's most Energy Friendly Microcontrollers. http://cdn.energymicro.com/dl/pdf/efm32_introduction_white_paper.pdf
8. Micro E (2011) Energy Friendly Radios: EFR4D2090 Datasheet. http://www.energymicro.com/draco. Confidential/Preliminary, Provided as Registered Copy
9. Polastre J, Szewczyk R, Culler D (2005) Telos: Enabling ultra-low power wireless research. In: Proceedings of the 4th international symposium on information processing in sensor networks (IPSN '05). IEEE Press, pp 364–369
10. Raman B, Chebrolu K (2008) Censor networks: a critique of "sensor networks" from a systems perspective. SIGCOMM Comput Commun Rev 38:75–78
11. Werner-Allen G, Lorincz K, Johnson J, Lees J, Welsh M (2006) Fidelity and yield in a volcano monitoring sensor network. In: Proceedings of the 7th symposium on operating systems design and implementation (OSDI '06). USENIX Association, pp 381–39
12. Saukh O, Sauter R, Meyer J, Marrón P (2008) Motefinder: a deployment tool for sensor networks. In: Proceedings of the workshop on real-world wireless sensor Networks (REALWSN)
13. Raman B, Chebrolu K, Madabhushi N, Gokhale DY, Valiveti PK, Jain D (2006) Implications of link range and (in)stability on sensor network architecture. In: Proceedings of the 1st

international workshop on wireless network ttestbeds, experimental evaluation & characterization (WiNTECH '06). ACM, pp 65–72
14. Nilsson M (2009) Directional antennas for wireless sensor networks. In: Proceedings of the 9th Scandinavian workshop on wireless adhoc networks (Adhoc)
15. Giorgetti G, Cidronali A, Gupta S, Manes G (2007) Exploiting low-cost directional antennas in 2.4 GHz IEEE 802.15.4 wireless sensor networks. In: Proceedings of the european conference on wireless technologies
16. Östrom E, Mottola L, Voigt T (2010) Evaluation of an electronically switched directional antenna for real-world low-power wireless networks. In: Proceedings of the 3rd international workshop on real-world wireless sensor networks (REALWSN)
17. Baccour N, Koubâa A, Jamaa MB, do Rosário D, Youssef H, Alves M, Becker LB (2011) Radiale: a framework for designing and assessing link quality estimators in wireless sensor networks. Ad Hoc Netw 9(7):1165–1185
18. Cerpa A, Busek N, Estrin D (2003) Scale: a tool for simple connectivity assessment in lossy environments. Tech. rep.
19. IEEE 802.15.4 Standard: http://standards.ieee.org/getieee802/download/802.15.4-2003.pdf (2003)
20. Chipcon cc2420: Data sheet. http://enaweb.eng.yale.edu/drupal/system/files/CC2420_Data_Sheet_1_4.pdf (2009)
21. Kotz D, Newport C, Elliott C (2003) The mistaken axioms of wireless-network research. Tech. rep., Dartmouth College
22. Zhao J, Govindan R (2003) Understanding packet delivery performance in dense wireless sensor networks. In: Proceedings of the 1st international conference on embedded networked sensor systems (SenSys '03). ACM, pp 1–13
23. Reijers N, Halkes G, Langendoen K (2004) Link layer measurements in sensor networks. In: Proceedings of the 1st IEEE international conference on mobile ad-hoc and sensor systems (MASS '04). IEEE Computer Society, pp. 24–27
24. Zuniga M, Krishnamachari B (2004) Analyzing the transitional region in low power wireless links. In: Proceedings of the 1st international conference on sensor and ad hoc communications and networks (SECON '04). IEEE Communications Society, pp 517–526
25. Zuniga M, Krishnamachari B (2007) An analysis of unreliability and asymmetry in low-power wireless links. ACM Trans Sen Netw 3(2):63–81
26. Mottola L, Picco GP, Ceriotti M, Gunǎ c, Murphy AL (2010) Not all wireless sensor networks are created equal: a comparative study on tunnels. ACM Trans Sen Netw 7:15:1–15:33
27. Zhou G, He T, Krishnamurthy S, Stankovic JA (2004) Impact of radio irregularity on wireless sensor networks. In: Proceedings of the 2nd international conference on mobile systems, Applications, and Services (MobiSys '04). ACM, pp 125–138
28. Ganesan D, Krishnamachari B, Woo A, Culler D, Estrin D, Wicker S (2002) Complex behavior at scale: an experimental study of low-power wireless sensor networks. Tech. rep.
29. Zhou G, He T, Stankovic JA, Abdelzaher T (2005) Rid: Radio interference detection in wireless sensor networks. In: Proceedings of the 24th annual joint conference of the IEEE computer and communications societies (INFOCOM '05). IEEE, pp 891–901
30. Zhou G, He T, Krishnamurthy S, Stankovic JA (2006) Models and solutions for radio irregularity in wireless sensor networks. ACM Trans Sen Netw 2(2):221–262. http://doi.acm.org/10.1145/1149283.1149287
31. Srinivasan K, Jain M, Choi JI, Azim T, Kim ES, Levis P, Krishnamachari B (2010) The κ factor: inferring protocol performance using inter-link reception correlation. In: Proceedings of the 16th annual international conference on mobile computing and networking (MobiCom '10). ACM, pp 317–328
32. Rappapport TS (2001) Wireless communications: principles and practice. Prentice Hall, Englewood Cliffs
33. Goldsmith A (2005) Wireless communications. Cambridge University Press, Cambridge
34. Cerpa A, Wong JL, Kuang L, Potkonjak M, Estrin D (2005) Statistical model of lossy links in wireless sensor networks. In: Proceedings of the 4th international symposium on information processing in sensor networks (IPSN '05). IEEE Press, pp 81–88

35. Liu H, Li J, Xie Z, Lin S, Whitehouse K, Stankovi, JA, Siu D (2010) Automatic and robust breadcrumb system deployment for indoor firefighter applications. In: Proceedings of the 8th international conference on mobile systems, applications, and services (MobiSys '10). ACM, pp 21–34
36. Cerpa A, Wong JL, Potkonjak M, Estrin D (2005) Temporal properties of low power wireless links: Modeling and implications on multi-hop routing. In: Proceedings of the 6th international symposium on mobile ad hoc networking and computing (MobiHoc '05). ACM, pp 414–425
37. Munir S, Lin S, Hoque E, Nirjon SMS, Stankovic JA, Whitehouse K (2010) Locationing burstiness for reliable communication and latency bound generation in wireless sensor networks. In: Proceedings of the 9th ACM/IEEE international conference on information processing in sensor networks (IPSN '10). ACM, pp 303–314
38. Brown J, McCarthy B, Roedig U, Voigt T, Sreenan CJ (2011) Burstprobe: debugging time-critical data delivery in wireless sensor networks. In: Proceedings of the 8th european conference on wireless sensor networks (EWSN '11). Springer-Verlag, pp 195–210
39. Srinivasan K, Kazandjieva MA, Agarwal S, Levis P (2008) The β-factor: measuring wireless link burstiness. In: Proceedings of the 6th international conference on embedded network sensor systems (SenSys '08). ACM, pp 29–42
40. Lin S, Zhang J, Zhou G, Gu L, Stankovic JA, He T (2006) Atpc: adaptive transmission power control for wireless sensor networks. In: Proceedings of the 4th international conference on embedded networked sensor systems (SenSys '06). ACM, pp 223–236
41. Lin S, Zhou G, Whitehouse K, Wu Y, Stankovic JA, He T (2009) Towards stable network performance in wireless sensor networks (rtss '09). In: Proceedings of the 30th IEEE real-time systems symposium. IEEE Computer Society, pp 227–237
42. Lal D, Manjeshwar A, Herrmann F (2003) Measurement and characterization of link quality metrics in energy constrained wireless sensor networks. In: Proceedings of the IEEE global telecommunications conference (Globecom '03). IEEE Communications Society, pp 446–452
43. Lee H, Cerpa A, Levis P (2007) Improving wireless simulation through noise modeling. In: Proceedings of the 6th international conference on information processing in sensor networks (IPSN '07). ACM, pp 21–30
44. Lymberopoulos D, Lindsey Q, Savvides A (2006) An empirical characterization of radio signal strength variability in 3-d ieee 802.15.4 networks using monopole antennas. In: Proceedings of the 7th european conference on wireless sensor networks (EWSN' 10). Springer, pp 326–341
45. Sikora A, Groza VF (2005) Coexistence of IEEE 802.15.4 with other systems in the 2.4 ghz-ISM-band. In: Proceedings of the IEEE conference on instrumentation and measurement technology (IMTC), pp 1786–1791
46. Petrova M, Wu L, Mähönen P, Riihijärvi J (2007) Interference measurements on performance degradation between colocated ieee 802.11g/n and ieee 802.15.4 networks. In: Proceedings of the international conference on networking (ICN), pp 93–98
47. Yang D, Xu Y, Gidlund M (2010) Coexistence of IEEE 802.15.4 based networks: a survey. In: Proceedings of the 36th annual conference on IEEE industrial electronics society (IECON). pp 2107–2113
48. Liang CJM, Priyantha NB, Liu J, Terzis A (2010) Surviving wi-fi interference in low power zigbee networks. In: Proceedings of the 8th ACM conference on embedded networked sensor systems (SenSys '10). ACM, pp 309–322
49. Boano CA, Voigt T, Noda C, Römer K, Zúñiga MA (2011) Jamlab: Augmenting sensornet testbeds with realistic and controlled interference generation. In: Proceedings of the 10th IEEE international conference on information processing in sensor networks (IPSN). pp 175–186
50. Zhou G, Stankovic JA, Son SH (2006) Crowded spectrum in wireless sensor networks. In: Proceedings of the 3rd workshop on embedded networked sensors (EmNets)
51. Son D, Krishnamachari B, Heidemann J (2006) Experimental study of concurrent transmission in wireless sensor networks. In: Proceedings of the 4th international conference on embedded networked sensor systems (SenSys '06). ACM, pp 237–250
52. Incel ÖD, Dulman S, Jansen P, Mullender S (2006) Multi-channel interference measurements for wireless sensor networks. In: Proceedings of the 31st IEEE international conference on communications (LCN). pp 694–701

53. Toscano E, Bello LL (2008) Cross-channel interference in IEEE 802.15.4 networks. In: Proceedings of the 7th international workshop on factory communication systems (WFCS). pp 139–148
54. Wu Y, Stankovic JA, He T, Lin S (2008) Realistic and efficient multi-channel communications in wireless sensor networks. In: Proceedings of the 27th IEEE international conference on computer communications (INFOCOM). pp 1193–1201
55. Xing G, Sha M, Huang J, Zhou G, Wang X, Liu S (2009) Multi-channel interference measurement and modeling in low-power wireless networks. In: Proceedings of the 30th IEEE international real-time systems symposium (RTSS). pp 248–257

Chapter 2
External Radio Interference

Abstract An important factor contributing to the degradation and variability of the link quality is radio interference. The increasingly crowded radio spectrum has triggered a vast array of research activities on interference mitigation techniques and on enhancing coexistence among electronic devices sharing the same or overlapping frequencies. This chapter gives an overview of the interference problem in low-power wireless sensor networks and provides a comprehensive survey on related literature, which covers experimentation, measurement, modelling, and mitigation of external radio interference. The aim is not to be exhaustive, but rather to accurately group and summarize existing solutions and their limitations, as well as to analyse the yet open challenges.

2.1 Introduction

The propagation of radio signals is affected by a plethora of variables, such as radio hardware, antenna irregularities, geometry and nature (static or mobile) of the environment, presence of obstacles responsible for shadowing or multipath fading, as well as the environmental conditions (e.g., temperature [1, 2]). These influences can lead to link unreliability and drastically vary the quality of a wireless link over time.

Another important factor that can vary the link quality and cause a degradation of communications is the presence of radio interference. Radio interference is indeed a serious challenge for wireless systems: it is caused by neighbouring devices operating concurrently in the same frequency band, disturbing each other by transmitting unwanted RF signals that play havoc with the desired ones. Interference is a severe problem especially for low-power wireless networks, as the presence of neighbouring devices transmitting at higher power may cause a significant degradation of the overall performance.

The primary outcome of interference is an increase in the packet loss rate, and it is in turn often followed by an increase in the network traffic due to retransmissions,

N. Baccour et al., *Radio Link Quality Estimation in Low-Power Wireless Networks*,
SpringerBriefs in Electrical and Computer Engineering,
DOI: 10.1007/978-3-319-00774-8_2,

as well as by a decrease in the performance and efficiency of the overall network. Experiences from several wireless sensor network deployments have shown that an unexpected increase of network traffic compared to the initial calculations may lead to an early battery depletion and/or deployment failure [3, 4].

Interference may also lead to unpredictable medium access contention times and high latencies. This is an important observation for low-power wireless networks used in safety-critical scenarios (e.g., industrial control and automation [5], health care [6], and high-confidence transportation systems [7]), where guaranteeing high packet delivery rates and limited delay bounds is necessary, and where unreliable connections cannot be tolerated.

On a network scale, interference can be both *internal* and *external*, and might affect the totality or only part of the nodes in the network.[1] *Internal interference* is the one generated by other sensor nodes operating in the same network, and it is typically mitigated by either proper placement of nodes and careful topology selection, or an appropriate MAC layer (e.g., making use of time diversity to avoid concurrent activities in the channel). *External interference* is caused by other wireless appliances operating in the same frequency range of the network of interest using other radio technologies. In the context of wireless networks composed of low-power sensor nodes, several devices operating at higher powers can be source of external interference (e.g., Wi-Fi access points and microwave ovens).

Because of its strong impact on the quality of wireless links, it is important to understand how interference affects the communications among wireless sensor nodes, and how to develop techniques that can properly mitigate its impact. Making wireless communications robust and reliable in the presence of interference is not an easy task. While internal interference can be minimized by means of a proper configuration of the network and protocol selection, the mitigation of external interference is often more complex for several reasons. Firstly, it is hardly possible to know in advance all potential sources of interference in a given environment and to predict their behaviour. Secondly, interference is often intermittent and highly dynamic, therefore it is difficult to create solutions that guarantee a reliable communication over time. Furthermore, fading due to multi-path propagation or shadowing from obstacles in the surroundings can affect the quality of wireless communications unpredictably, and things get even more erratic in the presence of interference because of the superposition of the unwanted radio signals [9].

In the next sections of this chapter we focus on wireless sensor networks, and we report on experimentation, measurement, modelling, and avoidance of external interference. We start reporting on the increasing congestion in the unlicensed frequency bands used by wireless sensor networks, as it makes their communications vulnerable to the interference generated by other wireless appliances.

[1] Some works also define *protocol interference* as the one occurring when multiple local protocols send conflicting commands to the radio transceiver [8].

2.2 Crowded Spectrum

Wireless communication technology has become increasingly popular in the last decades, because of the greater flexibility and remarkable reduction of costs for installation and maintenance compared to traditional wired solutions. This triggered a massive proliferation of wireless devices in our everyday life: the world of telecommunications and networking has experienced a large-scale revolution, and wireless systems have become ubiquitous, especially in residential and office buildings.

This proliferation has caused the radio spectrum to become a very expensive resource, therefore many standardized technologies operate in increasingly crowded and lightly regulated Industrial, Scientific and Medical (ISM) radio bands. The latter are freely-available portions of the radio spectrum internationally reserved for industrial, scientific and medical purposes other than communications.

When several technologies operate in the same ISM radio band, many devices concurrently share the same frequencies, and coexistence may become problematic. This applies especially to low-power wireless networks, as the presence of neighbouring devices transmitting at higher power may cause a significant degradation of the link quality, as well as a decrease in the packet delivery rates. Also wireless sensor networks compliant to the IEEE 802.15.4 standard are vulnerable to the external interference generated by neighbouring devices coexisting on the same frequencies, because the standard specifies operations in unlicensed and crowded ISM bands [10].

The IEEE 802.15.4 standard. The IEEE 802.15.4 protocol specifies the two lowest layers of the protocols stack for low-rate wireless personal area networks, namely the medium access control and physical layers. The physical layer (PHY) manages the physical RF transceiver, and provides the data transmission service: its main responsibilities are the data transmission and reception according to specific modulation techniques, as well as the channel frequency selection and management of energy and signal functions (e.g., LQI and energy detection). The first standard released in 2003 [11] specified only two PHY layers based on direct sequence spread spectrum (DSSS): the 868/915 PHY, employing binary phase-shift keying (BPSK) modulation, and the 2450 PHY, employing offset quadrature phase-shift keying (O-QPSK) modulation. Wireless sensor networks compliant to this standard were operating on one of three possible unlicensed ISM frequency bands:

- 868.0–868.6 MHz, available in Europe, one communication channel with center frequency $F_c = 868.3$ MHz;
- 902–928 MHz, available in North America, up to ten communication channels with center frequency $F_c = 906 + 2 \cdot (k - 1)$ MHz, for k = 1, 2, ..., 10;
- 2400–2483.5 MHz, available worldwide, up to sixteen communication channels with center frequency $F_c = 2405 + 5 \cdot (k - 11)$ MHz, for k = 11, 12, ..., 26.

The standardization process has been very active in the last decade, and important updates and amendments have been made to the first version of the IEEE 802.15.4 standard. A revision in 2006 [12] defined two optional 868/915 PHY layers employing a different modulation scheme, namely an 868/915 direct sequence spread spectrum PHY employing offset quadrature phase-shift keying (O-QPSK) modulation, and an 868/915 parallel sequence spread spectrum (PSSS) PHY employing a combination of binary keying and amplitude shift keying (ASK) modulation.

Because of the increasing congestion of the original three ISM band used, several amendments were defined to support new bands. The last version of the standard released in 2011 [13] encompasses the amendments defined by task groups IEEE 802.15.4.a/c/d and adds new physical layers, some of which make use of UltraWideBand (UWB) and Chirp Spread Spectrum (CSS) modulation techniques, as shown in Table 2.1.

UWB is a radio technology that employs high-bandwidth communications and uses a large portion of the radio spectrum, offering significant advantages with respect to robustness, energy consumption, and location accuracy compared to narrow-band DSSS. Because of the large bandwidth communications, UWB achieves a higher robustness against interference and fading, which makes this technology promising to achieve robust communications [14].

Despite the introduction of new ISM bands, the vast majority of wireless sensor networks deployed nowadays still makes use of the three original unlicensed ISM bands specified in the first edition of the IEEE 802.15.4 standard [11]. The main reason for this trend is essentially the availability of several off-the-shelf inexpensive hardware platforms developed in the beginning of the 21st century, such as the Moveiv TelosB and Sentilla Tmote Sky motes. Based on the available literature, we now briefly summarize the most prominent sources of interference in the three ISM bands specified in [11].

Interference in the 868/915 MHz ISM bands. Although the 868 and 915 MHz frequency bands are known to be relatively interference-free [15], several radio technologies have proven to cause significant problems to deployed wireless sensor networks. Barrenetxea et al. [16] have highlighted how cellular phones can be an interference source for sensor nodes operating in the 868 MHz band. The proximity of the European GSM band (that uses 890–915 MHz to send information from the mobile station to the base station and 935–960 MHz for the other direction), causes an impact on mote transmissions during the first seconds of an incoming call. Kusy et al. [17] have described the impact of in-band and out-of-band interference in the 900 MHz frequencies: in-band interference is caused by telemetry networks and cordless telephones, whereas mobile phones and pagers often cause out-of-band interference. Furthermore, several wireless devices marketed in Europe, including wireless domestic weather stations, car alarms, garage openers, and residential electronic alarms, use the 868 MHz frequency and are therefore potential sources of interference for wireless sensor networks operating in the 868/915 MHz ISM bands.

Interference in the 2.4 GHz ISM band. Most of the wireless sensor networks deployed nowadays use the 2.4 GHz frequencies, because they are available worldwide, and because several popular off-the-shelf sensor nodes embed radio transceivers operating in the 2450 PHY layer defined by the IEEE 802.15.4 standard (e.g., Moteiv TelosB, Crossbow MicaZ, and Sentilla Tmote Sky motes). However, to date, the 2.4 GHz is by far the most congested ISM band, because of the pervasiveness of devices operating in those frequencies, and their high transmission power. Sensor nodes must indeed compete with the communications of Wi-Fi (IEEE 802.11) and

Table 2.1 Frequency bands, modulation techniques, and data rates specified in the IEEE 802.15.4 standard [13]

PHY Layer (MHz)	Frequency Band (MHz)	Modulation	Bit Rate (kb/s)	Symbol Rate (ksymbol/s)	Symbols	802.15.4 Standard
780	779–787	O-QPSK	250	62.5	16-ary orthog.	2011 [13]
780	779–787	MPSK	250	62.5	16-ary orthog.	2011 [13]
868/915	868–868.6	BPSK	20	20	Binary	2003 [11]
	902–928		40	40		
868/915*	868–868.6	ASK	250	12.5	20-bit PSSS	2006 [12]
	902-928		250	50	5-bit PSSS	
868/915*	868–868.6	O-QPSK	100	25	16-ary orthog.	2006 [12]
	902–928		250	62.5	16-ary orthog.	
950	950–956	GFSK	100	100	Binary	2011 [13]
950	950–956	BPSK	20	20	Binary	2011 [13]
2450 DSSS	2400–2483.5	O-QPSK	250	62.5	16-ary orthog.	2003 [11]
2450 CSS*	2400–2483.5	DQCSK	250	167	64-ary orthog.	2011 [13]
			1000	167	8-ary orthog.	
UWB*	250–750 3244–4742 5944–10234	BPM-BPSK	Variable parameters			2011 [13]

*(optional)

Bluetooth (IEEE 802.15.1) devices, as well as with the noise generated by microwave ovens and other domestic appliances such as cordless phones, baby monitors, game controllers, presenters, and video-capture devices [10, 18–20].

As the 2.4 GHz band is, by far, the most crowded ISM band nowadays, we describe in the remainder of this section the characteristics of the most important sources of interference in this band, namely Wi-Fi devices, Bluetooth devices, and microwave ovens. We then conclude with a short discussion on the impact of adjacent channels interference from co-located IEEE 802.15.4 networks.

2.2.1 Coexistence Between IEEE 802.15.1/.15.4 Devices

The IEEE 802.15.1 (Bluetooth) standard specifies 79 channels, spaced 1 MHz, in the range 2402–2480 MHz, with center frequency $F_c = 2402 + k$, with $0 \leq k \leq 78$. Bluetooth uses the Frequency Hopping Spread Spectrum (FHSS) technology to combat interference and fading: it hops 1600 times per second, and therefore it remains at most 625 μs in the same channel. Given that only 79 channels are available, on average, one channel is used approximately 20 times each second: this makes interference generated by Bluetooth devices uniformly distributed across the whole 2.4 GHz band.

As of version 1.2, several Bluetooth stack implementations apply an Adaptive Frequency Hopping (AFH) mechanism to combat interference and fading, in which the hopping sequence is modified to avoid interfered channels. However, because the low-power sporadic communications of sensor nodes do not constitute a real threat compared to the communications between more powerful transmitters (e.g., Wi-Fi), Bluetooth devices will still make use of channels in which wireless sensor networks are operating.

The interference produced by IEEE 802.15.1 devices is, however, not so problematic for wireless sensor networks, because of the randomness and adaptiveness of Bluetooth's random frequency hopping [21]. Several experimental works have been carried out to study the impact of IEEE 802.15.1 communications on the reliability of sensornet transmissions. The packet loss rate of a wireless sensor network operating in the presence of Bluetooth interference is often between 3 % (as reported by Bertocco et al. [22] and Penna et al. [23]) and 5 % (as reported by Huo et al. [24]) up to a maximum of 9–10 % (as shown in the experimental results of Boano et al. [18] and Sikora and Groza [20]), hence not too critical.

2.2.2 Coexistence Between IEEE 802.11/.15.4 Devices

IEEE 802.11, better known as Wi-Fi, is a set of standards for wireless local area network (WLAN) communications. The standard divides the ISM bands into channels: the 2.4 GHz band (2400–2483.5 MHz), for example, is divided into up to 14 channels,[2] each of which has a bandwidth of 22 MHz. Channels are therefore partially overlapping, and there is only room for three orthogonal channels. The standard evolved significantly in the last decade (the first version was released in 1997), with data rates increasing from the original 2 Mbit/s to the 11 Mbit/s of 802.11b (1999), 54 Mbit/s of 802.11 g (2003), up to the 150 Mbit/s of 802.11n (2009); and it is still undergoing changes, with the new high-throughput 802.11ac protocol currently under development.

The coexistence between IEEE 802.11b/g/n and IEEE 802.15.4 devices represents a challenge for several reasons. Firstly, Wi-Fi devices are nowadays ubiquitous, especially in residential and office buildings where many Access Points (AP) are installed. Secondly, IEEE 802.11 devices operate at significantly higher power ($\approx$24 dBm) than traditional low-power sensor nodes. Thirdly, Wi-Fi channels have a bandwidth of 22 MHz and can therefore interfere with multiple IEEE 802.15.4 channels at the same time. Fourthly, IEEE 802.11 supports high-throughput transmissions that generate interference patterns that are difficult to predict, as they depend on factors such as the number of active users, their activities, the protocols they use (UDP or TCP), or the traffic conditions in the backbone.

Several works in the literature investigate the impact of IEEE 802.11 communications on the reliability of IEEE 802.15.4 transmissions, and show that wireless sensor networks suffer from high packet loss rates in the presence of Wi-Fi interference [19, 20, 22, 23, 25, 26]. Under certain conditions, also IEEE 802.11 devices can suffer from the interference of nearby wireless sensor networks. The communications between sensor nodes can indeed trigger a nearby Wi-Fi transmitter to back off: if this happens, only the header of IEEE 802.15.4 packets is typically corrupted [8, 26].

The actual packet loss rate that IEEE 802.15.4 networks experience in the presence of IEEE 802.11 transmissions depends on the Wi-Fi activity, as well as the location of the nodes. Boano et al. [18] have varied the Wi-Fi traffic pattern, showing that activities such as continuous radio streaming are not too critical for sensornet communications, as it results in approximately 15 % packet loss rate. On the contrary, activities such as video streaming ($\approx$30 % packet loss rate) and file transfers ($\approx$90 % packet loss rate) can destroy the majority of wireless sensor networks transmissions and cause long delays, drastically decreasing the performance of the network.

2.2.3 Coexistence Between Microwave Ovens and IEEE 802.15.4 Devices

Microwave ovens are a common kitchen appliance used to cook or warm food by exposing food to non-ionizing microwave radiations to make water and other polarized molecules oscillate, usually at a frequency of 2.45 GHz. Therefore, microwave ovens are a potential source of interference for wireless sensor networks operating in

[2] The availability of channels is regulated by each country, e.g., channel 14 is currently only available in Japan.

the 2.4 GHz spectrum. There are two main categories of microwave ovens: the ones designed for domestic use and the ones designed for commercial purposes, with the former being much more common [27].

The characteristics of the interference patterns emitted by domestic microwave ovens depend on the model; nevertheless all the ovens present the same basic properties. Firstly, with respect to the frequency spectrum, microwave ovens can potentially interfere on a large portion of IEEE 802.15.4 channels. The frequency of emitted microwaves depends on the sizes of the cavities of the magnetron, and varies also with changes in load impedance, supply current, and temperature of the tube. Also other factors, including the oven content, the amount of water in the food, and the position within the oven can affect the temperature of the magnetron [28]. Therefore, it is not possible to state with certainty which channel(s) will be mostly affected by the interference generated by microwave ovens. Secondly, with respect to the temporal behaviour, the noise generated by microwave ovens is rigorously periodical, as ovens continuously turn on and off according to the frequency of the AC supply line. The duration of a single period, called power cycle, hence mostly depends on the power grid frequency, but can also slightly vary depending on the oven model. Works in the literature report a power cycle of roughly 20 ms (at 50 Hz) or 16 ms (at 60 Hz) with an active period of at most 50 % of the power cycle [27, 29]. During the active period, the communications of low-power sensor nodes in proximity of a microwave oven are likely destroyed (because microwave ovens operate at up to $\approx$60 dBm), but during the inactive period several consecutive packet can be scheduled, as shown in [30]. The interference generated by microwave ovens can therefore be easily modelled as a deterministic sequence of interference pulses.

2.2.4 Coexistence Between IEEE 802.15.4 Devices Operating in Adjacent Channels

Several studies have highlighted that IEEE 802.15.4 channels in the 2.4 GHz ISM band are not orthogonal to each other, and hence wireless sensor networks operating on adjacent channels may interfere with each other [31–36].

Wu et al. [34] have shown that the number of orthogonal IEEE 802.15.4 channels in the 2.4 GHz ISM band is only eight, despite the actual number of channels with 5 MHz spacing available is sixteen. The authors carry out experiments using MicaZ nodes (that employ the CC2420 radio chip) and show that transmissions in adjacent channels decrease the packet reception rate, whereas transmissions generated at least two channels away from the one of interest do not harm the reception (that remains basically unaffected). The interference generated in the adjacent channel can decrease the packet reception rate as much as 50 % when using a low transmission power, leading to a potential disruption of connectivity that cannot be neglected. These results were confirmed by the experiments by Ahmed et al. [31]: the packet reception

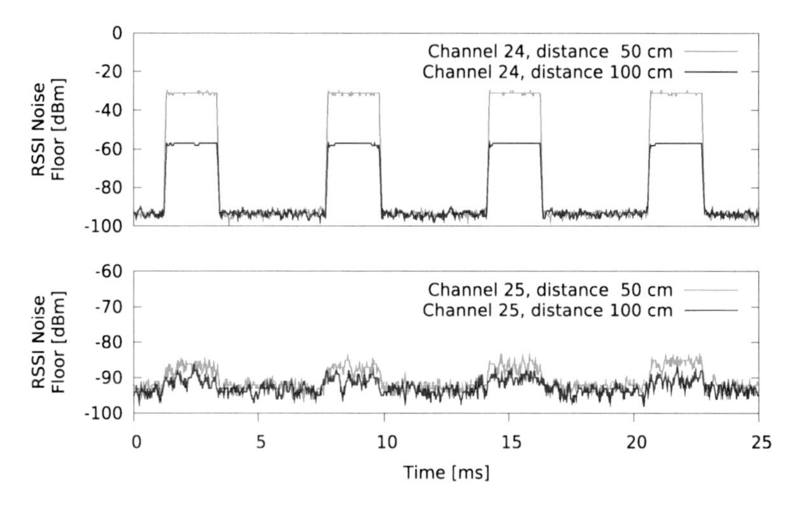

Fig. 2.1 IEEE 802.15.4 transmissions create noise on adjacent channels

rate decreases significantly in the presence of activities in adjacent IEEE 802.15.4 channels.

Figure 2.1 shows the impact of IEEE 802.15.4 packet transmissions on adjacent channels as measured by a Sentilla Tmote Sky node as follows. Two nodes $\mathcal{A}$ and $\mathcal{B}$ are placed in an interference-free environment at a given distance from each other (50 and 100 cm in two successive experiments). Node $\mathcal{A}$ listens first on channel 24 (center frequency at 2470 MHz) and later on channel 25 (center frequency 2475 MHz), recording the signal strength at the antenna pins. At the same time, node $\mathcal{B}$ continuously transmits packets at a rate of $\frac{1}{128}$ packets/s on channel 24. Figure 2.1 shows that the packet transmissions of node $\mathcal{B}$ on channel 24 generate also noise on channel 25 (bottom figure) that are high enough to affect communication in low-quality links. Hence, if node $\mathcal{A}$ would use channel 25 for its communications, it may experience packet loss due to the adjacent channel interference on channel 24.

2.3 Interference Measurement and Modeling

In the presence of external interference, the properties of a wireless channel can change unpredictably over time in both indoor and outdoor environments. Interference can be sporadic, causing only a temporary impact on communications, or persistent, causing a channel to experience heavy interference and become unavailable for long periods of time. Wireless sensor nodes may therefore need to adapt dynamically to changeable interference patterns and adjust their behaviour at runtime in order to maximize the reliability of their communications.

To adapt dynamically to various interference patterns, wireless sensor nodes firstly need to acquire a detailed understanding about the surrounding interference by means of accurate measurements. The latter must be carried out in a simple and energy efficient fashion, in order to meet the constrained capabilities of wireless sensor nodes. Secondly, efficient metrics to estimate the presence of interference need to be obtained from the measurements, in order to adapt dynamically to changing interference patterns, for example by carrying out a dynamic channel selection in multichannel protocols, or by ranking the available channels. Thirdly, there is a need to derive lightweight interference models that can be parametrized at runtime, in order to carry out a dynamic protocol selection or a dynamic adjustment of protocol parameters as soon as certain properties in the environment have changed.

In this section, we describe the most popular ways to accurately measure interference using off-the-shelf wireless sensor nodes (Sect. 2.3.1), and show how these measurements can be used to identify the active sources of interference in the surroundings (Sect. 2.3.2). We then describe simple interference models that can be implemented on resource-constrained wireless sensor nodes, and explain how they can be parametrized at runtime to achieve interference mitigation (Sect. 2.3.3).

2.3.1 Measuring Interference Using Sensor Nodes

The most common way to measure interference using wireless sensor nodes is the so called RF noise (floor) measurement, i.e., an estimate of the received signal power within the bandwidth of an IEEE 802.15.4 channel in absence of sensornet transmissions [37]. RF noise measurements are typically retrieved using the energy detection (ED) feature available in IEEE 802.15.4-compliant radios as part of a channel selection algorithm [11], and typically return an RSSI value that can be converted in dBm.

The key difference between RF noise measurements and traditional RSSI and LQI values is that the former can be obtained anytime, whereas the latter are generated only upon packet reception, and hence cannot describe the interference in a fine-grained way. Indeed, RSSI and LQI, as well as packet reception rate, can be used to derive the presence of interference and react accordingly (e.g., by switching channel when high packet losses arise [38]), but do not unequivocally identify and quantify the presence of interference in the environment. For example, the LQI describes the chip error rate for the first eight symbols following the SFD field, which obviously has a correlation with the amount of interference. Nevertheless, low LQI values may also result from unreliable links in absence of external interference. The same applies to the packet reception rate, as a low reception ratio may be caused by other factors than external interference, including routing issues, and software bugs.

Figure 2.2 shows the RSSI values returned by a Maxfor MTM-CM5000MSP sensor mote measuring RF noise in the presence of sensornet transmissions and external interference in the 2.4 GHz ISM band [39]. The mote employs the CC2420 radio and samples the RSSI at approximately 50 kHz, enough to capture the interference of

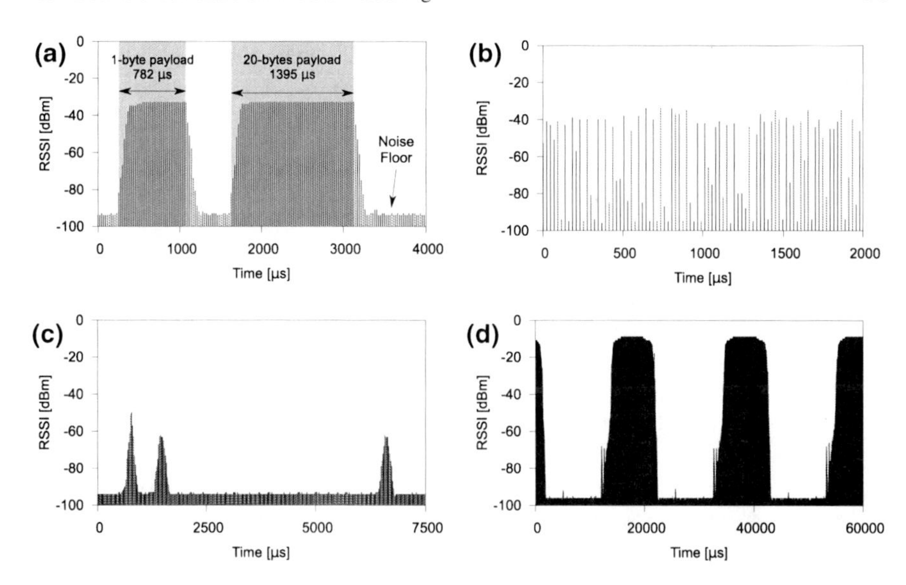

Fig. 2.2 RF noise values measured at a speed of 50 kHz using Maxfor MTM-CM5000MSP sensor motes operating in the 2.4 GHz ISM band [39]

common sources of interference such as microwave ovens, as well as several Bluetooth and Wi-Fi devices. In case interference is absent, the RSSI values are typically close to the sensitivity threshold of the radio and identify the noise floor (see Fig 2.2a). The latter has typically values between −100 and −95 dBm when using sensor nodes equipped with the CC2420 radio transceiver. As shown in Fig. 2.2, the interference generated from different devices produces different types of RSSI traces: one can easily identify the periodicity of microwave ovens, as well as the short and bursty transmissions of Wi-Fi devices.

Based on RF noise measurements, several works have proposed empirical metrics to derive the presence of interference. Musaloiou et al. [40] have measured the RF noise in each IEEE 802.15.4 channel at a rate of 20 samples/s in the presence of Wi-Fi activity and proposed three metrics that can be implemented on resource-constrained motes using off-the-shelf radios. The first metric is based on the cardinality of RSSI values, and counts the amount of unique RSSI values collected. Channels with high cardinality are likely to be rich of interference, as non-interfered channels have a small RSSI variation [18, 41]. The second metric is based on the maximum and mean RSSI value: channels with high levels of interference will record high maximum RSSI values as well as high average RSSI readings. The third metric counts the number of RSSI measurements above a given threshold: channels with the most measurements above such thresholds are considered to be highly interfered. This technique essentially consists in counting the amount of failed clear channel assessments carried out using a given RSSI threshold. The authors propose a threshold of −90 dBm, but this value may not be optimal because off-the-shelf sensors are typically uncalibrated [42].

Also Hauer et al. [25] suggest to predict the degradation of the link quality caused by interference using RF noise measurements. According to their experimental results gathered in an urban residential area, a substantial increase in RSSI noise floor values anticipates heavy packet loss, in particular when using very low transmission power levels.

Although very popular, RF noise measurements carried out using energy detection have several drawbacks:

1. Energy detection comes at a very high energy cost: the radio needs to be turned on in listening mode during the whole duration of the measurements.
2. A suitable RSSI sampling rate needs to be selected, a trade-off between energy-efficiency and accuracy. On the one hand, a low sampling rate may not capture the interference produced by devices transmitting at high frequency, e.g., Wi-Fi devices, and therefore assessing a medium as idle even though transmissions actually occurred between consecutive RSSI samples. On the other hand, a high sampling frequency corresponds to a significant energy expenditure that battery-powered sensor nodes may not be able to afford. Several works have tried to reach the limit of sensor nodes, and obtained RSSI sampling rates up to 62.5 kHz [41, 18]. This enables a quite accurate understanding of interference: Hauer et al. [41] exploit the RSSI profiles to estimate the bit error positions inside corrupted frames; Boano et al. [18] exploit the high-frequency samples to get a precise trace of interference for a later regeneration. However, even a resolution of 62.5 kHz is not sufficient to capture all possible interference sources: one can detect for example all IEEE 802.11b frames, but not all IEEE 802.11g/n frames.
3. Energy detection is a technique known to be brittle. RSSI readings have a low accuracy (± 6 dBm in the CC2420 radio, for example), and the readings may not be accurate in the presence of narrow unmodulated carriers because of saturation of the Intermediate Frequency (IF) amplifier chain, as highlighted in [18].
4. The sensor nodes often need to dedicate all their resources to the energy detection, and hence cannot receive and send packets or carry out other tasks, especially when sampling the RSSI with a very high frequency.

Because of the inefficiency of energy detection, a few works in the context of multichannel protocols do not make use of RF noise measurements to measure or detect the presence of interference. These works rely on either the packet reception rate [38, 43] or on the amount of failed transmissions and CCA failures [44, 45] to escape from congested channels. However, as mentioned above, these techniques have two main drawbacks: (i) they take effect only after the interference affected the communications within the sensor nodes, (ii) they do not unequivocally identify the presence of external interference.

2.3.2 Interferer Identification

An accurate measurement of interference can be used to identify which devices are generating the unwanted signals, a precious information that can be exploited to adapt packet transmissions and increase the robustness of communications. Interferers are typically classified based on RF noise measurements, but there are also a number of approaches based on the hardware estimators available in common IEEE 802.15.4-compliant radio transceivers such as RSSI and LQI (see Sect. 3.4).

Chowdhury and Akyildiz [30] describe a classification mechanism in which a sensor node actively scans channels in the 2.4 GHz ISM band to detect the interference generated by microwave ovens and IEEE 802.11b devices. In this approach, RF noise readings are used to obtain a spectral signature in all the available IEEE 802.15.4 channels. This signature is then used by the sensor nodes to derive the type of interferer by matching the observed spectral pattern with a stored reference shape. The knowledge about the type of interfering source is then exploited to construct an adaptive scheme for channel selection and scheduling of packets. Thanks to the correct classification, the authors show that the protocol can significantly reduce the packet loss rate in the presence of interference. However, as with other solutions exploiting RF noise measurements, this techniques may require a high amount of energy to scan all the available IEEE 802.15.4 channels, and cause the nodes in the network to be unreachable while performing energy detection.

Zacharias et al. [46] propose a classification algorithm that uses RSSI noise floor readings to monitor a single IEEE 802.15.4 channel in the 2.4 GHz ISM and understand which device is the source of interference. Compared to the approach by Chowdhury and Akyildiz, they also recognize Bluetooth interferers in addition to Wi-Fi devices and microwave ovens, but they do not exploit the knowledge to adapt packet transmissions. In particular, the authors exploit one second of RF noise readings (which results in approximately 11300 RSSI values) and classify the interferer depending on a given RSSI threshold, the channel usage, and the duration of interference over time. For example, their algorithm classifies as Wi-Fi interference all traces in which the usage of the channel is between 1 and 30 % and the maximum time of a clear channel is less than 100 ms (with an RSSI threshold of −85 dBm). Although the proposed duration of the scan (1 s) seems to be relatively energy efficient, it may not be enough to obtain a clear picture of the ongoing interference. Also, this approach cannot detect multiple sources at the same time and it is strongly dependent on empirical values that can vary depending on the calibration of different sensor nodes [42].

Differently from the previous two approaches, Hermans et al. [47] propose a method that is not based on energy detection. The authors combine the properties of LQI and RSSI during packet transmissions to investigate the feasibility of a lightweight interference detection and classification approach that only uses data that can be gathered during a sensor network's regular operation. Similarly to the work of Zacharias et al. [46], the authors try to differentiate between microwave ovens, Bluetooth and Wi-Fi devices, but without additional overhead due to scanning of the medium in absence of packet transmissions. In particular, they use LQI (that represents the chip error rate) to identify the cases in which a packet is received with a high LQI, but the packet fails the CRC check, which implies that channel conditions were good when reception started, but then deteriorated. The authors claim that such a sudden change in channel conditions can be observed when an interferer starts emission during packet reception. They further exploit the RSSI during packet reception to get a series of RSSI values for each received packet, containing about one sample per payload byte. Using this, the authors analyse the corruption of some 802.15.4 packets in the presence of different interferers and derive which parts of a packet have been corrupted, and use a supervised learning approach to train a classifier to assign each corrupted packet to a class representing the interfering device.

A work different in spirit from the above ones is that of Boers et al. [48, 49], which does not aim to identify the type of device, but rather the characteristic of interference. The authors distinguish interference with infrequent spikes, periodic spikes, periodic and frequent spikes in the RF noise measurements, as well as interference exhibiting a value constantly higher than the RSSI sensitivity threshold, or bimodal interference (such as the one produced by microwave ovens). In case a channel shows a mixture of these characteristics, the channel is classified according to the dominant pattern. The authors use decision trees as classifier, since, after training, the classification of new cases is simple, and can therefore meet the requirements of constrained wireless sensor nodes.

2.3.3 Modeling Interference

The creation of precise and lightweight models of common interference sources is not an easy task, because of the severe hardware limitations of wireless sensor nodes, and their reduced energy budget.

Very popular is the modelling of the channel occupancy as a two-state semi-Markov model, in which, at a given time instant, a channel is defined as busy if any interfering devices is transmitting packets and defined as idle otherwise [50]. The advantage of this simple model is that it can be easily parametrized at runtime using RF noise measurements: several works have defined a channel as busy when the RSSI values returned by RF noise measurements are above a certain threshold, and idle when the RSSI values are below such threshold [18, 51].

The parametrization of the lightweight two-state semi-Markov interference model at runtime has been used in the context of interference estimation and runtime adaptation of protocol parameters. Noda et al. [51] have presented a channel quality metric based on the availability of the channel over time, which quantifies spectrum usage. Distinctive feature of this metric is the ability to distinguish between a channel in which interference is bursty with large inactive periods and a channel that has very high frequency periodic interference with the same energy profile. As the first case is more favourable for having successful packet transmissions, the proposed metric ranks in a more favourable way channels with larger inactive periods or vacancies.

Using the two-state semi-Markov model of channel occupancy, Boano et al. [39] make use of RF noise measurements to measure the duration of the idle and busy instants, and compute the probability density function of idle and busy periods. Based on the duration of the longest busy period, the authors derive protocols parameters such that certain QoS requirements are met even in the presence of external interference.

The two-states model has also been extensively used with the purpose of generating interference. Boano et al. [18, 52] have implemented a bursty interferer model that sends continuous blocks of interference with uniformly distributed duration and spacing, which can be easily implemented on off-the-shelf sensors nodes to emulate, for example, the bursty transmission caused by Wi-Fi or Bluetooth devices. Interference follows continuous off/on periods and the transitions between the two states are specified by a Bernoulli random variable. A uniformly distributed random variable is further used to control the burstiness and duration of the interference. A second model, called semi-periodic interferer model, produces continuous blocks of interference as well, but the duration of the periods and their spacing have smaller variance, in order to emulate, for example, the type of interference generated by a sensornet performing periodic data collection.

In their follow-up work [18], the authors have followed a different approach: since the patterns generated by external interference differ substantially depending on the interfering source (as highlighted in Sect. 2.2), external interference was instead modelled per device. Because of its characteristics, microwave oven interference is simple to model, as it follows a deterministic on/off sequence. Hence, the interference is a function of three parameters τ, δ, ρ, where τ is the period of the signal, δ is the duty cycle (fraction of time the oven is "on") and ρ is the output power of the microwave, which determines the strength of the interference signal. The authors have used this model to generate a series of on/off signals resembling microwave oven interference using sensor motes and the CC2420 radio transceiver [18]. A similar model was used by Chowdhury and Akyildiz [30] for interference-aware scheduling of packets. To model Wi-Fi traffic in a simple and lightweight fashion, Boano et al. [18] resorted to an analytical model for saturated traffic sources, and derived models from empirical data for non-saturated traffic. In particular, for the latter, they denoted a random variable X as the *clear* time between two consecutive *busy* times, and obtained the probability mass function $p(x) = P_r\{X = x\}$ from the empirical sampling of the channel, where x is the time in number of slots (each slot is 1 ms). For saturated traffic, the authors have exploited the analytical model for the Distributed Coordination Function (DCF) mode of 802.11 proposed by Bianchi [53], and its simplification proposed by Garetto and Chiasserini [54].

2.4 Mitigating Interference

Several techniques have been proposed to tolerate external interference in wireless sensor networks [55]. In this section, we review related literature and propose a taxonomy that classifies existing interference mitigation techniques and aims to help

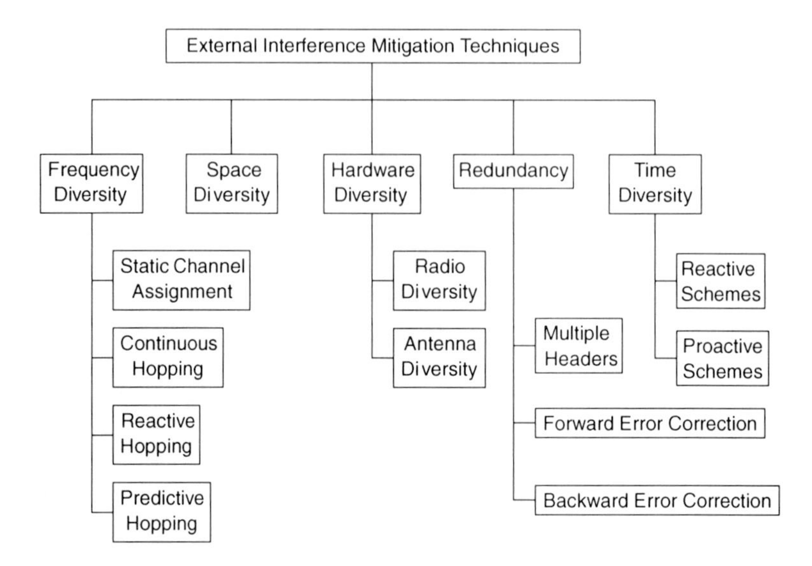

Fig. 2.3 Taxonomy for external interference mitigation techniques

protocol designers in identifying relevant mechanisms that make wireless sensor networks communications robust to external interference.

Most techniques exploit the availability of several radio channels in a given ISM band, for example by continuously hopping among channels over time. We group these applications in the frequency diversity class. Another popular interference mitigation technique consists in avoiding interference by routing packets through different links: we describe this solution in the space diversity class. Protocols deferring transmissions or scheduling packets in such a way to avoid interference are described in the time diversity class, whereas solutions exploiting multiple radios and antennas are grouped in the hardware diversity class. Finally, we define a class named redundancy, in which we group solutions that add redundancy to the transmitted information, such as the use of multiple headers and forward error correction techniques, as well as protocols making use of retransmissions. Figure 2.3 summarizes the proposed taxonomy of external interference mitigation techniques.

2.4.1 Frequency Diversity Solutions

One of the most common ways to mitigate external interference is to exploit the availability of several radio channels in a given ISM band.

Static channel assignment. The easiest way to pursue interference avoidance by exploiting multiple radio frequencies consists in statically assigning channels depending on the expected interference sources. This represents a very primordial way, that we name *static channel assignment*, in which the designer of the network essentially assigns each node to one or more fixed IEEE 802.15.4 channels that are supposedly interference-free for extended periods of time or throughout the lifetime of the network. In the easiest scenario, all nodes communicate on the same channel: this is the case for several real-world wireless sensor networks assigned to channel 26 in order to escape Wi-Fi interference in the 2.4 GHz ISM band [6, 8, 56]. This technique assumes that, in addition to Wi-Fi, no other interference source will ever interfere in that specific channel with the communications of the wireless sensor nodes, and hence it is highly unreliable. In more complex scenarios, involving dense wireless sensor networks covering large areas, different interference sources may be present throughout the network, and the quality of channels may differ from node to node. Several works have created multichannel protocols and statically assigned portions of the network to specific channels. However, this was mostly done with the intention of maximizing the bandwidth available for communications by increasing the number of channels for unicast transmissions [34, 57–61]. Hence, these multichannel solutions are not meant to provide any guarantee against external interference, and do not enhance coexistence among devices operating in the same frequency range.

Continuous hopping. Another way to pursue interference avoidance resembles the operations of the IEEE 802.15.1 standard (Bluetooth) and consists in continuously hopping among channels according to the same pseudo-random sequence. This technique, that we name *continuous hopping*, can be blind (i.e., nodes hop among all available channels) or adaptive, i.e., nodes carry out some form of blacklisting of undesirable channels where high traffic loads or excessive interference is present [62]. Examples of protocols adopting adaptive continuous hopping are the Time Synchronized Mesh Protocol (TSMP) [63], the Wireless-HART standard [64], the protocol developed by Du et al. [65] and Yoon et al. [66], and EM-MAC [45]. The latter is a typical example of adaptive continuous hopping: it uses a penalty system with channel blacklisting based on the results of the clear channel assessment (CCA) operation. A node switches among channels based on its pseudo-random channel schedule, except that, if the next pseudo-randomly chosen channel is on the node's channel blacklist, the node stays on its current channel.

Continuous hopping exploits the potential of frequency diversity and hence can reduce the impact of narrow-band interference and persistent multipath fading. Furthermore, channel hopping also ensures fairness among the chosen channels. However, channel hopping requires a tight time synchronization across the network in order to work properly. Also, the seed and the list of blacklisted channels needs to be shared in a reliable fashion, a critical operation in the presence of interference. It is important to note that in case of blind continuous channel hopping, the interference avoidance is only passive, i.e., by continuously hopping, a pair of nodes will sooner or later pick a good communication channel: this might not necessarily happen in a short time interval. Another drawback of continuous channel hopping is the additional energy required to continuously switch channels [67] on the long run, as well as the protocol overhead. Compared to protocols switching on demand, this represent a non-negligible burden. Also, adaptive continuous hopping protocols require to continuously update and spread the list of blacklisted channels, which may cause a significant energy consumption.

Reactive hopping. In order to avoid the burden of continuously switching the channel, several protocols switch or blacklist channels only once performance has degraded, e.g., in only a specific part of the network [68]. Indeed, these approaches continuously monitor the quality of the current channel and check whether it is satisfactory: if too much interference is detected, a frequency switch is carried out. We call these protocols *reactive hopping* protocols, as, to mitigate interference by frequency diversity, they first need to experience interference and a performance degradation. In this category fall protocols such as CoReDac [69], Chrisso [44], and ARCH [38]. The latter uses the expected number of transmissions (ETX) to monitor the quality of the link, and as soon as the ETX values collected in a given observation window exceed a certain threshold, a new channel is selected. The authors show that 15 minutes of observation are enough to predict channel reliability, and an interesting method is suggested to select the next channel. After blacklisting the current channel, ARCH hops to a new channel that is further away from the previous one. This has two benefits: on the one hand it avoids wideband interferers, on the other hand it avoids deep fades, as highlighted by Watteyne et al. [70].

The advantages of reactive protocols are, as mentioned above, the significant energy saving compared to hopping continuously and maintaining time synchronization among nodes. However, such protocols may not suit safety-critical systems, as they need to experience packet loss before performing a channel switch. Moreover, the switch is performed without knowledge about the stability of the other channels.

Proactive hopping. In order to overcome this problem, some other works try to avoid packet loss by predicting when the channel conditions will degrade and by hopping before this happens. We name protocols falling in this category *predictive hopping* protocols. This type of protocol typically uses channel quality estimation metrics to detect an early degradation of the channel [25, 40, 51]. However, because of the overhead required to continuously estimate the channel quality, no well-established proactive protocol has been developed yet. The work by Kerkez et al. [71] uses a sort of proactive strategy, as it keeps track of the quality of all channels by periodically forcing a channel switch. Another example is MuZi [72], in which as soon as the performance of a channel degrades, all the channels are scanned and a new reliable channel is selected. A fundamental role in the development of practical proactive protocols is played by proper interference metrics that can reliably assess the degradation of channels, as discussed in Sect. 2.3.1, as well as by an efficient link quality ranking algorithm [73].

2.4.2 *Space Diversity Solutions*

A solution widely investigated in the context of large and dense wireless sensor networks consists in avoiding interference by routing packets through different links [74]. Adaptive routing has been studied by several works that do not explicitly target the presence of external interference, but rather aim for an effective link estimation in order to achieve reliable communication.

Alizai et al. [75] proposed to apply a bursty routing extension to detect short-term reliable links. Their approach allows a routing protocol to forward packets over long-range bursty links in order to minimize the number of transmissions in the network. Liu and Cerpa [76] have developed a receiver-driven estimator based on a machine learning approach to predict the short temporal quality of a link. Their estimation is based on trained models that predict the link quality using both packet reception rate and other physical layer parameters, such as RSSI, SNR and LQI.

Gonga et al. [77] have carried out a comparison between multichannel communication and adaptive routing, in order to determine which one guarantees more reliable communications in the presence of external interference and high link dynamics. On the one hand, the authors have shown the good performance of adaptive routing in dense wireless sensor networks. The key reason behind this is the selection of good long-term stable links, which avoids low-quality links that may be more vulnerable to external interference. When external interference is present, one could indeed try to route a packet towards a closer node, so that the probability that a stronger signal corrupts the packet is smaller. On the other hand, in sparse networks, where the choice of forwarding links is rather limited, adaptive routing looses its flexibility, and multichannel solutions yield better performance in terms of both average end-to-end delay and reliability.

2.4.3 Hardware Diversity Solutions

In case an ISM band is highly congested and it is hardly possible to find a reliable channel for communications, or in case it is not possible to route a packet through a different link, a new strategy needs to be selected in order to mitigate external interference. Several works have made use of wireless sensor network platforms equipped with dual radios to communicate in multiple ISM bands. Other works have proposed the use of directional antennas or spatially separated antennas to achieve more robust communications even in the presence of interference. We group these solutions in the hardware diversity class, and describe them in more detail in the remainder of this section.

Radio diversity. Kusy et al. [17] have shown that radio transceivers operating at dual widely spaced radio frequencies and through spatially separated antennas offer robust communication, high link diversity, and better interference mitigation. Using dual radios, the authors have shown, through experiments, a significant improvement in the end-to-end delivery rates and network stability, at the price of a slight increase in energy cost compared to a single radio approach. This solution is rather effective, but it is however only feasible for wireless sensor network platforms equipped with dual radios, such as the BTnode, the Mulle node, and the Opal node.

Antenna diversity. Rehmani et al. [78] have envisioned the possibility to design and implement a software-defined intelligent antenna switching capability for wireless sensor nodes. More precisely, the authors have attached an inverted-F antenna to a TelosB mote in addition to the built-in antenna in order to achieve antenna diversity. Based on the wireless link condition, and in particular on physical layer measurements, the sensor node should then dynamically switch to the most appropriate antenna for communication.

Another options are dynamically steerable directional antennas, as shown experimentally by Giorgetti et al. [79]. Such antennas are able to dynamically control the gain as a function of direction, and, because of these properties, they can be very useful to increase the communication range and reducing the contention on the wireless medium. The authors have proposed a four-beam patch antenna and showed interference suppression from IEEE 802.11g systems. They have further discussed the use of the antenna as a form of angular diversity useful to cope with the variability of the radio signal. Other examples of directional antennas applied on wireless sensor nodes [80–82], in which a sensor node can concentrate the transmitted power towards the intended receiver dynamically.

2.4.4 Solutions Based on Redundancy

Several solutions exploit redundancy to mitigate the impact of external interference, such as the use of multiple headers, retransmissions, as well as forward error correction techniques. In the remainder of this section, we discuss those in detail.

Multiple headers. Liang et al. have experimentally shown that IEEE 802.11 transmitters can back off due to elevated channel energy when nearby IEEE 802.15.4 nodes start sending packets [8, 26]. When this happens, the IEEE 802.15.4 packet header is often corrupted, but the rest of the packet is still intact. Based on this observation, the authors have proposed the use of multi-headers to protect the IEEE 802.15.4 packets from the corruption generated by Wi-Fi interference. The authors have suggested that two additional headers represent a good trade-off between overhead and performance. It is important to notice that multi-headers are only effective in the so called symmetric region, i.e., when an IEEE 802.15.4 transmission is able to affect the behaviour of a IEEE 802.11 transmitter, because the bit errors occur mainly in the beginning of the packet. In contrast, in the asymmetric region, i.e., when the IEEE 802.15.4 signal is too weak to affect IEEE 802.11 behaviour, the bit errors are distributed across the packet in a uniform way.

Forward error correction techniques. In order to mitigate interference in the asymmetric region, Liang et al. [8, 26] have also proposed the use of forward error correction (FEC) techniques to recover from corrupted packets. FEC techniques use extra redundancy added to the original information frame to enable the correction of errors directly at the receiving node without the need for retransmission. When using forward error correction, the original message is encoded into a larger message by using an error correction code, which implies a longer time in which the radio is switched on, and a longer computation time for encoding and decoding the packet. The receiver than decodes the original message by applying the reverse transformation of the error correction code. The redundancy in the encoded message allows the receiver to recover the original message in the presence of a limited number of bit errors. The authors demonstrated that Reed-Solomon (RS) correcting codes perform well while recovering packets corrupted by the activities of IEEE 802.11 [26].

However, FEC techniques pose a trade-off between data recovery capacity and its inherent payload and computation overhead. Forward error correction indeed creates overhead both on the receiver and the transmitter, and therefore requires a significant amount of energy as well as powerful nodes. Liang et al. [26] have shown that the time required to encode an original 65-byte message into an RS-encoded message with 30-byte parity is approximately 36 ms, whereas the decoding of the message depends on the presence of errors and can vary between 100 and 200 ms.

Backward error correction techniques. An often used alternative to forward error correction is the use of acknowledgement (ACK) or negative-acknowledgement (NACK) packets to trigger a retransmission of the corrupted frames. This solution may not necessarily lead to a good result in the presence of external interference, as retransmitted packets are prone to corruption as well as the original packet. Furthermore, when sending ACK/NACK packets, one may increase the channel congestion and the energy consumption of the nodes.

In order to minimize the energy consumption required for retransmissions, Hauer et al. [41] have developed an Automatic Repeat reQuest (ARQ) scheme that minimizes the amount of data that the sender needs to retransmit. In their scheme, a receiving node records the RSSI of the received packet during reception at high frequency, and tries to estimate the position of the error within the packet. The RSSI-based recovery mechanisms is effective also in the presence of external interference, because collisions of frames with the transmissions generated by other devices such as IEEE 802.11 or IEEE 802.15.4 can be detected through an increase in the RSSI profile, which would otherwise be very stable (typically ± 1 dBm), as one can see in Fig. 2.4. The authors then propose an RSSI-based recovery mechanism, in which the receiving node triggers only the retransmission of the damaged portion of the packet, which implies a significant amount of energy in case of packets with relatively long data payloads.

Exploiting the stability of RSSI over time in absence of interference, Boano et al. [39] have proposed a novel protocol that uses jamming signals instead of message transmissions as the last step of a packet handshake between two nodes. This permits the two nodes to reliably agree on whether a packet was correctly received even in the presence of external interference. Agreement is indeed an issue in the presence of external interference, as ACK packets may be lost as well as original packets, and there is no way to guarantee the actual reception of a given packet. If two sensor nodes need to agree, for example, on a new time slot or frequency channel, message loss caused by external interference may break agreement in two different ways: none of the nodes use the new information (time slot, channel) and stick with the previous assignment, or—even worse—some nodes use the new information and some do not, leading to reduced performance and failures [39]. To get around this problem, the authors have proposed a jamming-based agreement protocol that con-

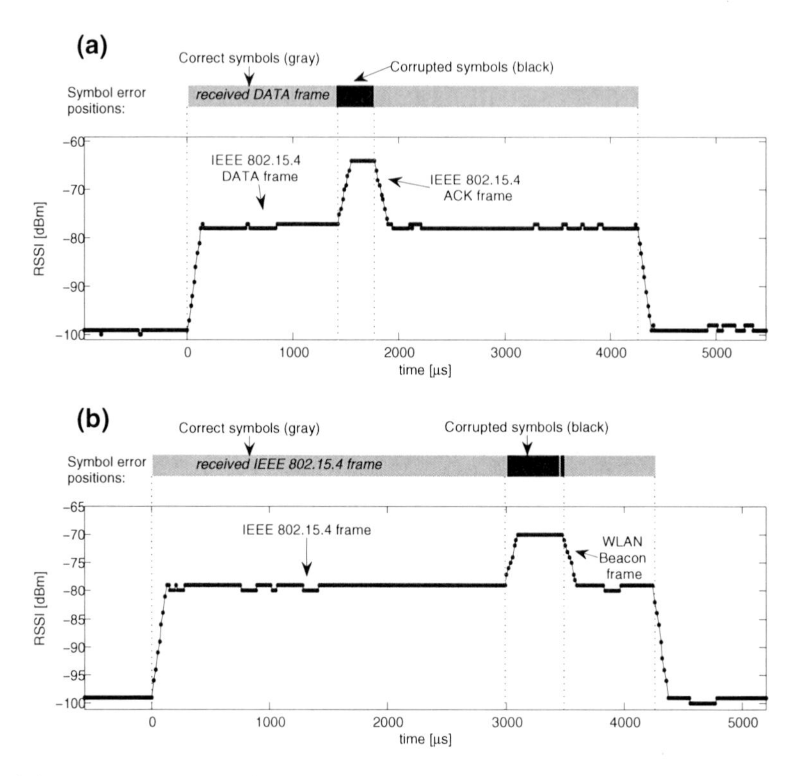

Fig. 2.4 Error positions and RSSI profiles of an IEEE 802.15.4 frame (133-byte PHY Protocol Data Unit) colliding with an IEEE 802.15.4 packet (*top*) and IEEE 802.11 beacon frame (*bottom*). The RSSI profile is measured on a Tmote Sky mote placed at less than 5 m from the interferer [41]

sists in a three-way handshake in which the last acknowledgement packet is sent in the form of a jamming signal. The transmission of a jamming signal has the property of being easily recognizable even in the presence of interference. As shown in Fig. 2.5, a jamming sequence results in a stable RSSI value above the sensitivity threshold of the radio, whereas in the presence of additional external interference, the RSSI register will return the maximum of the jamming signal and the interference signal due to the co-channel rejection properties of the radio. However, typical interference sources—in contrast to a jamming signal—do not produce continuous interference for long periods of time, rather they alternate between idle and busy. Therefore, by sending a jamming signal lasting long enough (i.e., longer than the longest busy period of the interference signal), one can unequivocally detect the absence of the jamming signal, by checking if any of the RSSI samples equals the sensitivity threshold of the radio. The authors have further shown that by carefully selecting a proper jamming time window, one can guarantee the identification of the ACK despite the presence of external interference, which makes this approach suitable for safety-critical systems.

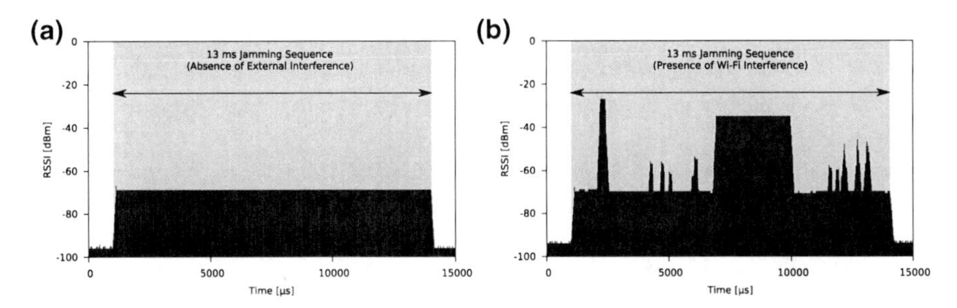

Fig. 2.5 RSSI values recorded during the transmission of a jamming sequence without external interference (**a**), and with external Wi-Fi interference (**b**) [39]

2.4.5 Time Diversity Solutions

Another class of external interference mitigation techniques is time diversity, which consists in either deferring transmissions, or scheduling them in such a way to avoid interference.

Reactive schemes. A basic way to mitigate interference is to defer transmission until interference clears, using for example the Carrier Sense Multiple Access with Collision Avoidance (CSMA-CA) technique. Those solutions, that we name *reactive*, react to the given interference pattern by adjusting protocol or parameter settings accordingly. In this context, the role of congestion back-off and Clear Channel Assessement (CCA) has been studied extensively in the presence of external interference. Boano et al. [52] have investigated the selection of a suitable congestion back-off scheme when using CCA and detecting a busy channel. They also investigate protocols or parameter settings that enable potentially more handshakes in case some fail due to interference, and extend X-MAC, showing an improved robustness to interference.

An important role is also played by the CCA threshold. Yuan et al. [83] have proposed a decentralized approach in which sensor nodes adaptively and distributively adjust their CCA thresholds in the presence of external heavy interference. The authors show that this approach substantially reduce the amount of discarded packets due to channel access failures, and hence increases the performance of sensornet protocols under interference.

Bertocco et al. [84] study the performance of different CCA modes in the presence of in-channel wide-band additive white gaussian noise (AWGN). Similarly, Petrova et al. [19] investigate the three CCA modes defined by the IEEE 802.15.4 PHY standard (energy above threshold, carrier sense only, and carrier sense with energy above threshold). They have observed that dynamic CCA thresholds can improve the performance of sensornet communications both in the overlapping and non-overlapping channels with IEEE 802.11n.

Proactive schemes. Another class of protocols is the one in which the sensor node tries not to defer transmissions, but rather to schedule them in a way to avoid the interference of other devices. An example from this class, that we name *proactive schemes*, is the scheduling of packets proposed by Chowdury and Akyildiz [30]. In their protocol, the authors analyze the cases in which microwave ovens and Wi-Fi device are operating, and propose a scheme in which the sensor nodes transmit whenever the channel is predicted to be free based on the Wi-Fi traffic or microwave oven duty cycle. In the case of microwave oven interference, the sensor nodes can align their own sleep cycles with the duty cycle of the microwave oven, and synchronize their transmissions with the beginning of the off-time. In the case of Wi-Fi transmissions, the sensor nodes exploit the detection of Short Inter-Frame Space (SIFS) and Distributed Inter-Frame Space (DIFS). As also highlighted by Liang et al. [26], the peaked power pulses emitted by the sensor nodes interrupt the DIFS carrier sense and force a back-off among the contending Wi-Fi devices, leaving the channel free for the sensor nodes to complete their transmissions.

2.5 Experimenting with Interference

As interference can severely affect the reliability of wireless communications, there is a strong need for understanding the performance of existing sensornet protocols under interference, as well as designing and validating novel protocols that can deliver high and stable performance despite changing interference patterns. Furthermore, testing of the correct functionality of the system prior deployment is of fundamental importance in wireless sensor networks: several reports from real-world deployments have highlighted how the lack of testing can lead to a partial or complete system failure [3, 4].

An accurate validation and testing requires a proper infrastructure, in which realistic interference patterns can be created in an easily controllable, inexpensive, accurate, and repeatable way. This is in practice very difficult to obtain, and especially when it comes to radio interference and wireless communications, experimentation can be frustrating and may require a large amount of time and resources.

On the one hand, wireless propagation is extremely complex and dependent on a plethora of variables, such as radio hardware, antenna irregularities, geometry and nature (static or mobile) of the environment, presence of obstacles responsible for shadowing or multipath fading, as well as the environmental conditions (e.g., temperature [2]). These influences can only be modelled to a limited extent in simulators, and cannot be easily controlled when experimenting using real-hardware.

On the other hand, radio interference can be produced by several devices, and the generated patterns can be highly diverse, causing a given protocol to perform differently in different scenarios. Hence, one would need to verify several different possibilities and control several variables, which is in practice very difficult to obtain. Indeed, all possible scenarios cannot be exhaustively verified using real-hardware or simulators, as their number would be prohibitively large [85].

For these reasons, experimentation with radio interference, whether carried out in simulation, in a laboratory testbed, or in a real-world deployment, can be frustrating and time-consuming, and needs to address several key aspects, such as accuracy and repeatability, that we analyse in the remainder of this section.

2.5.1 Requirements

We now summarize a list of key properties that experiments (carried out both in simulation and using real-hardware) involving radio interference should satisfy.

Realism and accuracy. When testing the reliability and robustness of a protocol or an application against external interference in a systematic fashion, one needs to set up realistic and credible experiments. The interference patterns used in the experiments must be accurate and be a representative set of how interference appears in reality. Having a device that is permanently interfering for long periods of time would not represent a realistic scenario, as it hardly occurs in practice (interference is instead typically bursty).

Device diversity. Experiencing packet loss due the presence of external interference is rather common nowadays, as it is typically enough to operate a sensor network in the presence of active Wi-Fi transmissions. In order to achieve a complete investigation, however, it is important to use different interference sources. A given protocol may be resilient to the periodic interference generated by microwave ovens, but may suffer the randomness of Bluetooth or the bursty nature of Wi-Fi transmissions, hence testing the protocol using only one interfering device is often not optimal. One should also make use of several heterogeneous devices at the same time, especially when testing protocols that exploit frequency diversity. For example, a multichannel protocol that blacklists congested channels would provide very good results in the presence of only one interfering device operating permanently on a given channel. However, the use of several wide-band devices operating concurrently using different frequencies may cause the protocol to perform poorly.

Spatial diversity. Because of the complexity of wireless propagation, and the intrinsic properties of low-power hardware such as antenna irregularities (see Chap. 1), it is also important to vary the location of both interfering sources and network nodes. Very often one aims to generate the "worse case" and hence places the interfering device very close to the sensor motes, so to block their communications. This is often referred to as binary interference [18]: if a device is active, it automatically interferes the operations of a given sensor node. However, this may not necessarily lead to the worse case setting, as the most challenging scenarios may often be the ones in which interference affects the communications between a pair of nodes only intermittently, or when it only affects parts of the nodes in the network.

Temporal diversity. Different devices can interfere in different ways, and it is therefore important to try different patterns for each interference source. For example, in the case of a Wi-Fi device, if one varies the user activity, the transport protocol, or the packet size and other low-level parameters, one may obtain significantly different interference patterns that may lead to a different performance. Similarly, it is important to vary the usage pattern of a given device, as they often differ on the long-term.

As an example, the amount of traffic generated by a Wi-Fi station typically varies significantly from day to night and from weekday to weekends.

Scalability and controllability. The infrastructure for interference generation should support as many network nodes and interfering devices as possible, without requiring an excessive amount of time and resources, or additional costs. It is also very important to be able to easily control the interfering devices, as well as to ensure that the interference patterns that are generated correspond to the expected ones. This may not be an easy task in practice, as several devices often need to be set manually. For example, the activation of several microwave ovens in a large-scale network would be hardly feasible if the devices cannot be programmed remotely.

Repeatability. Being able to repeat an experiment under the same interference patterns is one of the most challenging aspects of experimenting with interference, especially when comparing the performance of different protocols on real hardware. On the one hand, wireless propagation is extremely complex and may be affected by several external factors that are usually unknown to the experimenter. On the other hand, most experiments are carried out in "RF jungles" such as office environments and residential buildings [86], where one does not usually have control on the background noise level or the presence of people, nor one can make sure that the experimental setup did not change between consecutive experiments, as discussed in Sect. 2.5.3.

2.5.2 *Experimenting Using Simulators*

Even though simulation environments offer high levels of scalability, controllability, and repeatability, they still lack realistic and accurate behaviour, especially when it comes to simulating wireless communications and interference. Although several solutions have been proposed in the literature to make simulation of wireless communications and interference more accurate [87–89], there are still several fundamental and implementation issues that affect the accuracy of simulators.

A fundamental issue is represented by mathematical models, such as the unit disk model, and the log-normal shadowing model, which do not capture the actual behaviour of wireless sensor nodes accurately [90]. This problem neutralizes in a sense the simplicity with which the experimenter can relocate network nodes and interfering devices in a simulation environment, as the inaccuracy of the propagation model may not be true to reality. Furthermore, the parametrization of the simulation models is still a major challenge to make simulation experiments realistic, as their choice is often rather arbitrary. Trace-based models such as the one offered by TOSSIM have several limitations, including the fact that the noise is established separately and randomly for each of the nodes, failing in representing link burstiness [88].

There are also a number of implementation issues. On the one hand, the repeatability across different simulators may be limited due to the usage of random number generators, as highlighted by Garg et al. [90]. On the other hand, simulators often come with a limited amount of models, and implementing a new solution (e.g., extending an existing simulator with impulsive interference in wireless sensor networks [91]) may not be trivial. Especially if one needs to simulate the behaviour of different devices generating interference at the same time, designing and implementing new models may also require a substantial amount of time.

As the creation of models that accurately reflect reality is a rather difficult task, some works propose to augment existing simulation tools with the playback of realistic interference traces [92]. An example of a simulator providing such capabilities is COOJA [93]: traces recorded using off-the-shelf sensor motes can be incorporated directly into the simulation environment, improving the level of realism.

Despite the above limitations of existing simulators, a substantial number of experiments and evaluations regarding the impact of external radio interference have been carried out using simulation. Nethi et al. [94] have implemented a multichannel protocol to avoid interference and simulated its performance under interference using the ns-2 simulator, but did not provide details about the interference generation. Yuan et al. [83] have proposed a decentralized approach that adjusts CCA thresholds of IEEE 802.15.4 nodes in the presence of heavy interference, and validated it using OPNET using two scenarios involving IEEE 802.11b nodes. Several parameters of the simulation (transmission power, receiver sensitivity, data rate, CCA threshold, payload size) are given, along with a description of the position of the interfering nodes. OPNET has also been used by Shin et al. [95] to investigate the coexistence of ZigBee and Bluetooth devices exploiting the Suitetooth package, and several parameters of the simulation (line-of-sight distance, path loss exponent, payload size, transmission power) are reported. Voigt et al. [69] evaluate a multichannel protocol in COOJA by adding disturber nodes generating interference that prevents the use of selected channels. Similarly, Iyer et al. [96] use jamming nodes on specific portions of the network in TOSSIM.

Due to the aforementioned limitations of existing simulators, some works have extended existing simulators with new interference patterns or implemented a new custom simulator. Bertocco et al. have extended the OMNeT++ simulator and supported the analysis of interference between IEEE 802.11b and IEEE 802.15.4-based networks [97]. Dominicis et al. have used this simulator to investigate coexistence issues when using the WirelessHART protocol [98]. Chowdhury et al. have written a custom C++ simulator to simulate the transmissions of IEEE 802.11b within a wireless sensor network, and varied the number of Wi-Fi nodes between 10 and 20, each of which generated packets at different rates [30]. Boers et al. have extended a SIDE-based emulator by adding the ability to define scripted external impulsive interference by creating a user-specified configuration, a new node type and running threads producing the specified interference [91].

2.5.3 Experimenting Using Real Hardware

Experiments on real-hardware, whether carried out in a real-world deployment or in a laboratory testbed, offer a higher level of accuracy and realism compared to simulation, but they do not scale well, and they often come at a very high cost.

Existing sensornet testbeds lack capabilities for interference generation, and upgrading them with additional heterogeneous devices in order to introduce interference sources is a costly, inflexible, and labor-intensive operation. One cannot easily place bulky equipment such as Wi-Fi access points and microwave ovens into a laboratory testbed, or, even worse, bring several devices to the deployment location, and ensure a stable power supply. Also, changes in the experimental setup, including the relocation or addition of interference sources, may need to be done manually. Changes in the behaviour of the interfering sources require manual or remote activation and programming of each device, which creates a significant overhead (especially in the case of unconventional equipment such as microwave ovens).

To avoid this problem, a number of sensornet testbeds are equipped with heterogeneous devices in order to enable the generation of different types of interference. One example is EasiTest [99], in which high-speed multi-radio nodes and low-speed single-radio nodes have been used to study the co-existence problem between IEEE 802.11 and IEEE 802.15.4 devices. However, the number and the size of such heterogeneous testbeds is still rather limited.

In the last years, many researchers have also proposed the use of Software Defined Radio (SDR) devices, such as USRP and WARP, to generate dynamic interference patterns. This choice is mainly driven by the easy reconfiguration and adaptivity of software defined radios. Following this idea, Sanchez et al. [100] have envisioned a novel testbed federation incorporating SDR devices, which would facilitate recording and playback of interference patterns. However, the cost of SDR hardware is still very high, and hence this approach does not scale to large testbeds.

A solution that was proposed in order to avoid the need of heterogeneous devices and speed up the setup time of experiments in an inexpensive way consists in using sensor motes as interfering devices [18, 69, 101, 102]. Although sensor motes can only interfere on a single channel and with limited transmission power, and hence the accuracy of the generated patterns is rather low, the clear advantage of this approach are the limited setup time, and the use of sensor nodes (no additional hardware required).

A big concern when experimenting using real-hardware experiments is repeatability. Burchfield et al. [86] have described the environments in which most wireless experiments are carried out as real "RF jungles", as way too many assumptions are made on the environment, and there is hence a high risk of misinterpreting the data obtained from such experiments. For example, it is not necessarily true that experiments conducted at night are interference-free. In the same way, one can hardly make sure that, when generating specific interference patterns, no other interference source is present in the environment, as wireless propagation is affected by several external factors that are usually unknown to the experimenter and that can severely bias experimental results. Indeed, very often, sensornet testbeds are located in office and residential buildings rich of activities of other wireless devices [103], and only very few studies have been carried out in special environments, e.g., anechoic chambers [47, 104] or shielded Faraday cages [30].

Another problem comes from the fact that real-world interference cannot be easily repeated. Gnawali et al. have highlighted that evaluating protocols by running them one at a time on real-hardware is not optimal since no experiment is absolutely repeatable [105], and this applies especially to experiments involving wireless systems, as wireless propagation depends on a myriad of factors. For example, experiments exploiting "ambient interference" surrounding the wireless sensor network testbed may not be a suitable option to compare several protocols or applications under realistic interference, as the interference patterns are not fully controllable and cannot be recreated precisely. This may not enable a fair comparison among different protocols.

Based on the available literature, we now classify the existing works studying external interference experimentally on real-hardware in four different categories, in which the experiments are carried out by:

- exploiting noisy environments;
- generating specific interference patterns exploiting existing equipment;
- generating specific interference patterns using a custom setup;
- using sensor nodes to generate interference.

Exploiting noisy environments. Several experiments exploit the "ambient interference" surrounding the wireless sensor network to evaluate their protocols or applications under realistic interference patterns.

Gonga et al. [77] run experiments in which the IEEE 802.11 access points co-located with the sensor network in an office environment act as sources for narrow band interference. The authors also report the presence of microwave ovens and people moving throughout the day, which makes the experiment more realistic due to the shadowing and multipath fading being introduced. Open office environments have also been used in several other experiments [65, 106–110].

Also university campuses have hosted several experiments, such as the ones by Zhou et al. [111] and Hauer et al. [41]. In the latter, the authors run their experiments on publicly accessible indoor sensor network testbeds deployed in university buildings (TWIST and MoteLab) surrounded by Wi-Fi access points. Noda et al. [51] have analyzed the bursty distribution of interference in a university library, and reported heavy traffic from co-located Wi-Fi networks.

Finally, Sha et al. [112] have analysed the spectrum usage in residential environments and collected several observations used to derive a new frequency-adaptive MAC protocol [38].

Exploiting noisy environments has two key advantages: the realism of real-world interference patterns, and the rather low-cost and effort required to set up the experiments, as there is no need to explicitly generate interference. This comes at the price of a complete uncontrollability and non-repeatability of the experiments: one cannot have full knowledge of the devices that are actually interfering, nor can one differentiate their impact.

Generating specific interference patterns exploiting existing equipment. In several works, the authors generate specific interference patterns exploiting existing infrastructures.

Dutta et al. [113] evaluate their protocol on a university testbed with 94 TelosB nodes, and actively verify the performance while "a file transfer is in progress using a nearby 802.11 access point". No information is reported on the protocol used and on the size of IEEE 802.11 packets. Similarly, Moeller et al. [114] perform experiments on a 40-motes indoor wireless sensor network testbed, and operate two 802.11 radios on a given channel, transmitting UDP packets of size 890 bytes. Also Lao et al. [115] and Kang et al. [116] exploit existing Wi-Fi access points in the proximity of an indoor testbed and download, from a laptop, a large amount of files from a server using the FTP protocol. Liang et al. [26] evaluated their protocol on a 57-node testbed and tested its performance using the interference originating from a co-located IEEE 802.11g testbed on the same building floor. Full details on the 802.11g testbed and the devices downloading are given. Musaloiou and Terzis [40] run six consecutive UDP transfers at different rates and report the impact on the Zigbee network, along with a detailed map of the testbed and the location of the access points. Rohde et al. [117] performed a measurement to quantify interference in a 2-story residential building. The authors used an iPhone in active connection with a Wi-Fi access point. Zhao et al. [99] directly program the existing infrastructure of Easitest and generate 1480-byte-long IEEE 802.11 UDP packets to study the coexistence with co-located wireless sensor networks.

Exploiting existing infrastructure to specifically generate certain interference patterns is a popular approach, and offers several advantages and disadvantages. On the one hand, the time required to set up the experiments is minimal, since one can exploit the existing infrastructure. On the other hand, the experimenter is limited by the amount, location, and type of the devices, and often cannot make sure that no other interference source was active in addition to the one being generated. Interestingly, most of the works following this approach consist in "continuous file transfer using Wi-Fi", but without any further documentation or justification.

Generating specific interference patterns using a custom setup. The most popular way to experiment with interference seems, by far, to be the creation of a custom experimental setup. The latter is typically small-scale and involves a small amount of sensor nodes and the interfering device of interest (typically one or two units).

Examples of this approach are the works by Won et al. [68], that created an ad-hoc network of two 802.11b devices and run a file transfer. The same approach was used by Hauer et al. [25], Sikora and Groza [20], Ahmed et al. [31, 118, 119], Tang et al. [45] (that uses the iperf network testing tool instead of a file transfer), Ansari et al. [120], Petrova et al. [19], Hossian et al. [121], Huang et al. [122], Xu et al. [72], Shuaib et al. [123], and Jeong et al. [124].

Several studies experiment with multiple interfering devices, e.g., IEEE 802.11 and Bluetooth interference. This is the case for the studies of Penna et al. [23],

Bertocco et al. [22], and Arkoulis et al. [125]. Similar experiments were also carried out by Hou et al. [24] and Boano et al. [18], but in this case, also microwave ovens were used to generate interference. A distinctive feature of the experiments of Boano et al. [18] with Wi-Fi devices is the differentiation of the interference depending on the user activity. In their experiments, the authors present different results depending on the activity of the user on the machine (file transfer, video streaming, radio streaming).

Common features of experiments carried out using a custom setup are the rather detailed description of the experimental facility that, however, typically uses only a few (often just one) interfering device in a single configuration of speed, power and protocol. Especially in studies involving more advanced equipment, such as signal generators, and software-defined radios [126, 127], the setup is limited to a couple of nodes, but offers high degrees of controllability.

Using sensor nodes to generate interference. Several works [69, 101, 102] use sensor motes as jammers and continuously transmit random packets, introducing noise into the frequencies of interest and evaluating the impact of such interference on several protocols. Although the transmission of packets is not fully controllable (e.g., inter-packet times [128]) and does not resemble the interference produced by devices using other technologies, the clear advantage of this approach are the limited setup time, and the use of sensor nodes (no additional hardware required). This kind of solution can be very useful if the experimenter is primarily interested in binary interference generation, i.e., in a setup in which the interfering nodes either blocks the communication of the motes in their surroundings by emitting a strong-enough interference signal, or by not interfering at all.

Boano et al. [18] have enhanced this methodology and proposed to augment existing sensornet testbeds with JamLab, a low-cost infrastructure for the creation or playback of realistic and repeatable interference patterns. With JamLab, either a fraction of the existing nodes in a testbed are used to record and playback interfer-

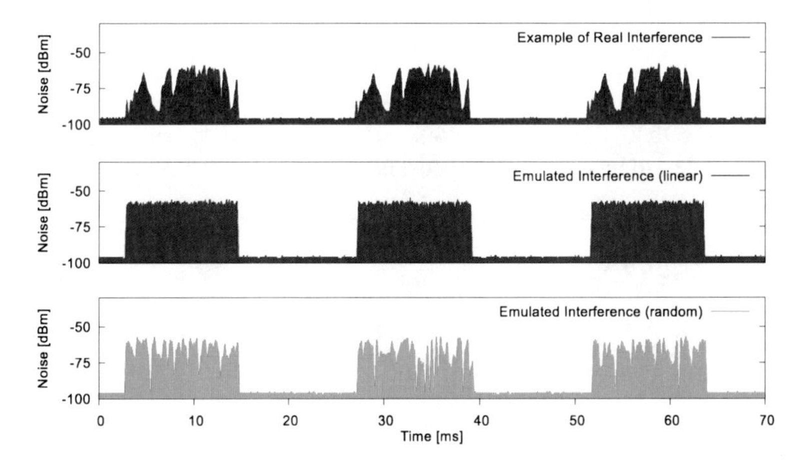

Fig. 2.6 Emulation of microwave oven interference (*top*) with fixed (*middle*) and random power (*bottom*) using JamLab [18]

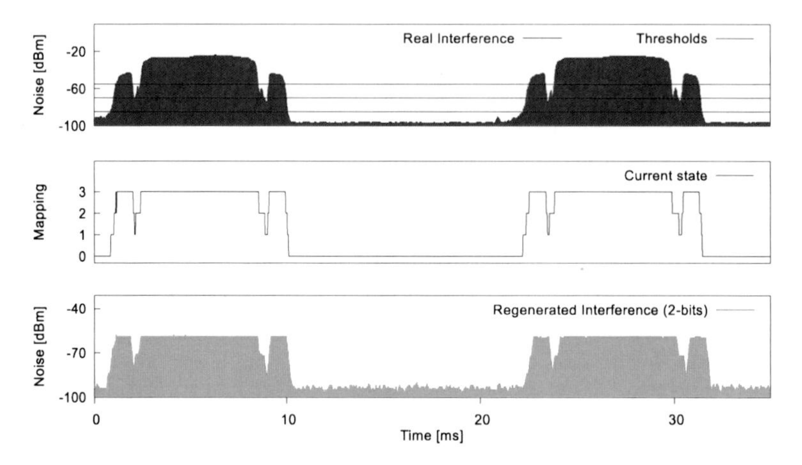

Fig. 2.7 Regenerated interference of a microwave oven using JamLab [18]

ence patterns, or a few additional motes are placed in the testbed area; hence the installation overhead is minimal. The nodes selected to generate interference can have two modes of operation: emulation, where a simplified model is used to generate interference patterns that resemble those generated by a specific appliance (such as a WiFi device or a microwave oven); and regeneration, where each interfering node autonomously samples the interference in the environment, compresses and stores it locally, and regenerates the recorded patterns later. The latter mode offers the possibility to record realistic interference patterns at the deployment site, and bring them back to the laboratory testbed for a more detailed study of their impact. Figures 2.6 and 2.7 show an example of emulated and regenerated interference using JamLab [18]. On the one hand, the interference is an accurate representation of a given model and fully repeatable (given that no other uncontrollable source of interference is present in the environment), as it repeats over time continuously. On the other hand, the hardware used to generate the interference are off-the-shelf sensor motes, and therefore one can only emulate the behaviour of other devices with all the limitations of the sensor nodes, e.g., limited transmission power, operations on a single IEEE 802.15.4 channel. Several works have exploited JamLab to generate interference in existing sensornet testbeds, including [44, 52, 129–131].

2.5.4 Observations

Based on a careful analysis of existing published work in the area, we now illustrate our observations regarding experimentation with interference.

1. **Most experiments are carried out using real-hardware**. Because of the inaccuracy of existing simulators and because of the complexity of deriving reliable and precise models of different interference sources, the majority of experiments are carried out using real devices.

2. **Table experiments are very popular**. Most experiments involving real hardware are carried out using customized setups involving only few sensor nodes and one interfering device "on top of a table". This is probably due to the necessity of selecting and controlling the interference patterns, and due to the fact that interference would be difficult to control on a large scale. On the other hand, a small-scale setup cannot capture situations in which different portions of a large network experience different interference patterns. Also in most "table experiments" the transmissions/noise generated by the interfering devices are typically strong enough to destroy the sensornet communication, given the short distances. It is hence not possible to capture cases in which the interference destroys only some of the packets sent by sensor nodes.
3. **Lack of frequency diversity**. It is perhaps not surprising that most experiments are carried out using sensor nodes operating in the 2.4 GHz ISM band, given the high number of devices and technologies sharing these frequencies. However, very few studies are carried out in other frequency bands [15], and few studies are known about potential interference patterns caused by medical and RFID devices, as well as cellular phones in other frequencies.
4. **Lack of device diversity**. Most of the experiments target a specific interference source, the latter being often Wi-Fi, with a few experiments carried out under Bluetooth and microwave oven interference. It follows that several interference sources have still not been considered, and that external interference in wireless sensor networks is often conceptually referred to as Wi-Fi interference, or that Wi-Fi interference implicitly captures the worst-case of external interference.
5. **Lack of temporal diversity**. Apart from being carried out using Wi-Fi devices, the majority of experiments involving external interference often makes use of "continuous file transfers". This choice is probably driven by the intention of generating a worse-case scenario with high levels of interference, which is typically the case of heavy file transfers. However, this is not necessarily a good choice. A multichannel protocol in the presence of a heavily congested channel would likely blacklist the channel and not experience any problem, compared to a multichannel protocol in the presence of intermittent interference among different channels. Also, the choice of file transfer often comes without a justification, and no further configurations of the interfering device are tested.
6. **Insufficient descriptions**. A fundamental task of scientists is the meticulous description of the setup and the surrounding environment in scientific papers. Although the typical size of a manuscript limits the space available for describing the setup of an experiment, some information is sometimes indispensable for a deep understanding of the results. For example, a fundamental piece of information is often the actual congestion of the medium during the experiments, which helps in understanding the packet loss rates. Only a few works, such as [65, 112, 114] provide the spectral traces of the available channels, and it is often difficult to understand the reasons for a certain performance. Without resorting to additional devices to capture the spectrum, one may just run one of the existing interference estimation metrics, such as [51], and report its results together with the performance of protocols, as done in [39]. Alternatively, one may use the Expected

Network Delivery [132], a metric that quantifies the delivery performance that a collection protocol can be expected to achieve given the network topology.

7. **Monitoring background activity**. A recurring assumption is the absence of unwanted signals during an experiment. Even the smallest and most controlled setup may however experience unwanted interfering sources, often unknown to the experimenter. It is a common practice in several works to use spectrum analyzers to verify the actual absence of unwanted interfering sources. Very popular is also the use of portable cheaper spectrum analyzers such as the Wi-Spy [25, 26, 32, 41, 65, 86, 107, 112].
8. **Comparability among experiments is difficult**. Despite several experiments are carried out using Wi-Fi and file transfers, it is difficult to compare them. Firstly, different experiments often have a different setup. Secondly, devices can be configured in several ways (for example, Wi-Fi access points) and the configuration information is rarely reported. Furthermore, the interference generated by file transfers using Wi-Fi devices often depends on the congestion of the backbone or on the actual distance from the access point. Therefore, results should be taken with a grain of salt, and their comparison is rather difficult.

Although experimentation with interference evolved significantly in the last years, these observations show that further research is necessary. Experiments involving sensor nodes and interfering devices "on top of a table" still represent the most common experimental setup, and there are very few studies that analyse the performance of a given protocol under interference generated using different devices and varying interference patterns.

Also, the congestion of the 2.4 GHz band has now been extensively studied, and there is a need to understand the performance of communication protocols in specific settings. For example, in hospital and clinical structures, it is important to make sure that sensor networks do not interfere with medical equipment and vice-versa, as alarm procedures must be triggered immediately when monitoring life-threatening deteriorations in vital signs of hospitalized patients.

Furthermore, a large number of protocols have been developed and proposed to mitigate the impact of interference, but there is no comprehensive study comparing their performance under different interference patterns. Experiments similar to [52] would significantly help in understanding which approach (i.e., frequency hopping, forward or backward error correction, routing through different links) offers the best performance in the presence of a given interference pattern.

2.6 Conclusion

Radio interference is receiving increasing attention in wireless sensor network research. As more and more wireless devices are being deployed, there is a strong need to increase the robustness of communications carried out in unlicensed frequencies, as they are vulnerable to the interference generated by other wireless appliances.

In order to cope with the massive proliferation of wireless devices in our everyday life, new standards are being introduced to increase the number of unlicensed frequencies that can be used for communication. Recent amendments, for example, have standardized the use of the UWB technology for wireless sensor networks, as it offers excellent interference immunity and low complexity at rather low costs. New frequency bands will be released in the near future to mitigate coexistence problems in existing bands (e.g., in the IEEE 802.11ac standard, Wi-Fi devices will not operate in the 2.4 GHz band anymore, and therefore a big portion of potential interfering devices will move to other frequencies). However, the increasing rate in which wireless devices are being deployed hints that there will still be several devices sharing the same frequencies, and that hence will need to coexist. This requires the design of protocols that guarantee reliable and robust communications among wireless sensor networks, especially for real-time and safety-critical applications that can hardly tolerate high packet loss rates and long latencies.

In the last decade, the research community has actively tackled the problem, and provided several solutions aimed to mitigate external interference in wireless sensor networks. The use of multichannel protocols, radio diversity, redundancy in all its forms (e.g., forward error correction techniques, multiple packet headers) was shown to increase significantly the robustness of communications in the presence of external interference.

However, the dynamism of interference requires solutions that can adapt at runtime to the changing interference patterns. In their renowned hitch-hiker's guide to successful wireless sensor network deployments, Barrenetxea et al. [16] recommend to make sure that the radio frequency used by the sensor nodes is not already in use. This task is hard to fulfil, as interference changes dynamically over time, and therefore there is a need for lightweight and efficient solutions that adapt to interference at runtime. The research community still needs to come up with self-learning solutions that can efficiently adapt the behaviour of sensor nodes at runtime, such as a dynamic protocol selection, or the dynamic adjustment of model parameters. The main obstacle towards this goal is currently the inefficiency of energy detection, as the radio transceiver needs to be turned on in listening mode for extended periods of time.

Recent advances in cognitive radio technologies have highlighted the possibility to apply dynamic spectrum access techniques in order to get access to less congested spectrum [133]. Cognitive radio capable sensor nodes can adapt to varying channel conditions and adapt their parameters at runtime, in order to increase the transmission efficiency and save energy. The research community has hence started to explore this promising paradigm, in order to adopt cognitive radio capabilities in wireless sensor networks in the near future.

References

1. Boano CA, Brown J, He Z, Roedig U, Voigt T (2009) Low-power radio communication in industrial outdoor deployments: the impact of weather conditions and ATEX-compliance. In: Proceedings of the 1st international conference on sensor networks applications, experimentation and logistics (SensAppeal), pp 159–176
2. Boano CA, Brown J, Tsiftes N, Roedig U, Voigt T (2010) The impact of temperature on outdoor industrial sensornet applications. IEEE Trans Industr Inform 6(3):451–459
3. Beutel J, Römer K, Ringwald M, Woehrle M (2009) Deployment techniques for sensor networks. In: Ferrari G (ed) Sensor networks. Springer, Berlin, pp 219–248
4. Langendoen K, Baggio A, Visser O (2006) Murphy loves potatoes: experiences from a pilot sensor network deployment in precision agriculture. In: Proceedings of the 14th international workshop on parallel and distributed real-time systems (WPDRTS), pp 174–181
5. GINSENG: Performance control in wireless sensor networks. http://www.ict-ginseng.eu/
6. Chipara O, Lu C, Bailey TC, Roman GC (2010) Reliable clinical monitoring using wireless sensor networks: experiences in a step-down hospital unit. In: Proceedings of the 8th ACM international conference on embedded networked sensor systems (SenSys), pp 155–168
7. Jeong J (2009) Wireless sensor networking for intelligent transportation systems. Ph.D. thesis, University of Minnesota, MN, USA
8. Liang CJM (2011) Interference characterization and mitigation in large-scale wireless sensor networks. Ph.D. thesis, John Hopkins University, Baltimore
9. Son D, Krishnamachar B, Heidemann J (2006) Experimental study of concurrent transmission in wireless sensor networks. In: Proceedings of the 4th international conference on embedded networked sensor systems (SenSys '06). ACM, New York, pp 237–250
10. Zhou G, Stankovic JA, Son SH (2006) Crowded spectrum in wireless sensor networks. In: Proceedings of the 3rd workshop on embedded networked sensors (EmNets)
11. IEEE 802.15.4 Working Group (2003) IEEE Standard for Local and Metropolitan Area Networks—Part 15.4: Wireless Medium Access Control (MAC) and Physical Layer (PHY) Specifications for Low-Rate Wireless Personal Area Networks (WPANs), IEEE std 802.15.4-2003 edn
12. IEEE 802.15.4 Working Group (2006) IEEE Standard for Local and Metropolitan Area Networks—Part 15.4: Wireless Medium Access Control (MAC) and Physical Layer (PHY) Specifications for Low-Rate Wireless Personal Area Networks (WPANs), IEEE std 802.15.4-2006 edn
13. IEEE 802.15.4 Working Group (2011) IEEE Standard for Local and Metropolitan Area Networks—Part 15.4: Low-Rate Wireless Personal Area Networks (LR-WPANs), IEEE std 802.15.4-2011 edn
14. Zhang J, Orlik PV, Sahinoglu Z, Molisch AF, Kinney P (2009) UWB systems for wireless sensor networks. Proc IEEE 97(2):313–331
15. Woehrle M, Bor M, Langendoen KG (2012) 868 MHz: a noiseless environment, but no free lunch for protocol design. In: Proceedings of the 9th international conference on networked sensing systems (INSS)
16. Barrenetxea G, Ingelrest F, Schaefer G, Vetterli M (2008) The Hitchhiker's guide to successful wireless sensor network deployments. In: Proceedings of the 6th ACM international conference on embedded networked sensor systems (SenSys), pp 43–56
17. Kusy B, Richter C, Hu W, Afanasyev M, Jurdak R, Brünig M, Abbott D, Huynh C, Ostry D (2011) Radio diversity for reliable communication in WSNs. In: Proceedings of the 10th IEEE international conference on information processing in sensor networks (IPSN), pp 270–281
18. Boano CA, Voigt T, Noda C, Römer K, Zúñiga MA (2011) Jamlab: augmenting sensornet testbeds with realistic and controlled interference generation. In: Proceedings of the 10th IEEE international conference on information processing in sensor networks (IPSN), pp 175–186
19. Petrova M, Wu L, Mähönen P, Riihijärvi J (2007) Interference measurements on performance degradation between colocated ieee 802.11g/n and ieee 802.15.4 networks. In: Proceedings of the international conference on networking (ICN), pp 93–98

20. Sikora A, Groza VF (2005) Coexistence of IEEE 802.15.4 with other systems in the 2.4 GHz-ISM-band. In: Proceedings of the IEEE conference on instrumentation and measurement technology (IMTC), pp 1786–1791
21. Farahani S (2008) ZigBee wireless networks and transceivers. Elsevier Inc., Amsterdam
22. Bertocco M, Gamba G, Sona A (2008) Is CSMA/CA really efficient against interference in a wireless control system? an experimental answer. In: Proceedings of the 13th IEEE international conference on emerging technologies and factory automation (ETFA), pp 885–892
23. Penna F, Pastrone C, Spirito M, Garello R (2009) Measurement-based analysis of spectrum sensing in adaptive WSNs under Wi-Fi and bluetooth interference. In: Proceedings of the 69th IEEE vehicular technology conference (VTC), pp 1–5
24. Huo H, Xu Y, Bilen CC, Zhang H (2009) Coexistence issues of 2.4 GHz sensor networks with other RF devices at home. In: Proceedings of the 3rd international conference on sensor technologies and applications (SENSORCOMM), pp 200–205
25. Hauer JH, Handziski V, Wolisz A (2009) Experimental study of the impact of WLAN interference on IEEE 802.15.4 body area networks. In: Proceedings of the 6th European conference on wireless sensor networks (EWSN), pp 17–32
26. Liang CJM, Priyantha NB, Liu J, Terzis A (2010) Surviving Wi-Fi interference in low power zigbee networks. In: Proceedings of the 8th ACM conference on embedded networked sensor systems (SenSys '10). ACM, New York, pp 309–322
27. Kamerman A, Erkocevic N (1997) Microwave oven interference on wireless LANs operating in the 2.4 GHz ISM band. In: Proceedings of the 8th IEEE international symposium on personal, indoor and mobile radio communications (PIRMC), vol. 3, pp. 1221–1227
28. Vollmer M (2004) Physics of the microwave oven. Phys Education 39:74–81
29. Taher TM, Misurac MJ, LoCicero JL, Ucci DR (2008) Microwave oven signal modeling. In: Proceedings of the IEEE wireless communications and networking conference (WCNC), pp 1235–1238
30. Chowdhury KR, Akyildiz IF (2009) Interferer classification, channel selection and transmission adaptation for wireless sensor networks. In: Proceedings of the IEEE international conference on communications (ICC), pp 1–5
31. Ahmed N, Kanhere S, Jha S (2009) Multi-channel interference in wireless sensor networks. In: Proceedings of the 8th IEEE international conference on information processing in sensor networks (IPSN), poster session, pp 367–368
32. Bello LL, Toscano E (2009) Coexistence issues of multiple co-located IEEE 802.15.4/zigbee networks running on adjacent radio channels in industrial environments. IEEE Trans Ind Inform 5:157–167
33. Incel ÖD, Dulman S, Jansen P, Mullender S (2006) Multi-channel interference measurements for wireless sensor networks. In: Proceedings of the 31st IEEE international conference on communications (LCN), pp 694–701
34. Wu Y, Stankovic JA, He T, Lin S (2008) Realistic and efficient multi-channel communications in wireless sensor networks. In: Proceedings of the 27th IEEE international conference on computer communications (INFOCOM), pp 1193–1201
35. Xing G, Sha M, Huang J, Zhou G, Wang X, Liu S (2009) Multi-channel interference measurement and modeling in low-power wireless networks. In: Proceedings of the 30th IEEE international real-time systems symposium (RTSS), pp. 248–257
36. Xu X, Lao J, Zhang Q (2010) Design of non-orthogonal multi-channel sensor networks. In: Proceedings of the 30th IEEE international conference on distributed computing systems (ICDCS), pp 358–367
37. Srinivasan K, Dutta P, Tavakoli A, Levis P (2010) An empirical study of low-power wireless. ACM Trans Sens Netw 6:1–49
38. Sha M, Hackmann G, Lu C (2011) ARCH: practical channel hopping for reliable home-area sensor networks. In: Proceedings of the 17th IEEE international real-time and embedded technology and applications symposium (RTAS), pp 305–315

39. Boano CA, Zúñiga MA, Römer K, Voigt T (2012) JAG: reliable and predictable wireless agreement under external radio interference. In: Proceedings of the 33rd IEEE international real-time systems symposium (RTSS), pp 315–326
40. Musaloiu-ER, Terzis A (2007) Minimising the effect of WiFi interference in 802.15.4 wireless sensor networks. Int J Sens Netw (IJSNet) 3(1):43–54
41. Hauer JH, Willig A, Wolisz A (2010) Mitigating the effects of RF interference through RSSI-based error recovery. In: Proceedings of the 7th European conference on wireless sensor networks (EWSN), pp 224–239
42. Chen Y, Terzis A (2010) On the mechanisms and effects of calibrating RSSI measurements for 802.15.4 radios. In: Proceedings of the 7th European conference on wireless sensor networks (EWSN). LNCS 5970, pp 272–288
43. Doddavenkatappa M, Chan MC, Leong B (2011) Improving link quality by exploiting channel diversity in wireless sensor networks. In: Proceedings of the IEEE 32nd real-time systems symposium (RTSS), pp 159–169
44. Iyer V, Woehrle M, Langendoen K (2011) Chrysso: a multi-channel approach to mitigate external interference. In: Proceedings of the 8th IEEE communications society conference on sensor, mesh, and ad hoc communications and networks (SECON)
45. Tang L, Sun Y, Gurewitz O, Johnson DB (2011) EM-MAC: a dynamic multichannel energy-efficient MAC protocol for wireless sensor networks. In: Proceedings of the 12th ACM international symposium on mobile ad hoc networking and computing (MobiHoc), pp 23:1–23:11
46. Zacharias S, Newe T, O'Keeffe S, Lewis E (2012) Identifying sources of interference in RSSI traces of a single IEEE 802.15.4 channel. In: Proceedings of the 8th international conference on wireless and mobile communications (ICWMC)
47. Hermans F, Rensfelt O, Voigt T, Ngai E, Larzon LA, Gunningberg P (2013) SoNIC: classifying interference in 802.15.4 sensor networks. In: Proceedings of the 12th ACM/IEEE international conference on information processing in sensor networks (IPSN)
48. Boers NM, Nikolaidis I, Gburzynski P (2010) Patterns in the RSSI traces from an indoor urban environment. In: Proceedings of the 15th international workshop on computer aided modeling, analysis and design of communication links and networks (CAMAD), pp 61–65
49. Boers NM, Nikolaidis I, Gburzynski P (2012) Sampling and classifying interference patterns in a wireless sensor network. ACM Trans Sens Netw (TOSN) 9(1):1–19
50. Stabellini L, Zander J (2010) Energy-efficient detection of intermittent interference in wireless sensor networks. Int J Sens Netw (IJSNET) 8(1):27–40
51. Noda C, Prabh S, Alves M, Boano CA, Voigt T (2011) Quantifying the channel quality for interference-aware wireless sensor networks. ACM SIGBED Rev 8(4):43–48
52. Boano CA, Voigt T, Tsiftes N, Mottola L, Römer K, Zúñiga MA (2010) Making sensornet MAC protocols robust against interference. In: Proceedings of the 7th European conference on wireless sensor networks (EWSN). LNCS 5970, pp 272–288
53. Bianchi G (2000) Performance analysis of the IEEE 802.11 distributed coordination function. IEEE J Sel Areas Commun 18(3):535–547
54. Garetto M, Chiasserini CF (2005) Performance analysis of 802.11 WLANs under sporadic traffic. In: Proceedings of the 4th IFIP-TC6 networking conference (NETWORKING), Waterloo, Canada
55. Yang D, Xu Y, Gidlund M (2011) Wireless coexistence between IEEE 802.11- and IEEE 802.15.4-based networks: a survey. Int J Distrib Sens Netw (IJDSN) 2011:17pp
56. Angelopoulos CM, Nikoletseas S, Theofanopoulos GC (2011) A smart system for garden watering using wireless sensor networks. In: Proceedings of the 9th ACM international symposium on mobility management and wireless access (MobiWac), pp 167–170
57. Incel ÖD, Jansen P, Mullender S (2011) MC-LMAC: a multi-channel MAC protocol for wireless sensor networks. J Ad Hoc Netw 9:73–94
58. Kim Y, Shin H, Cha H (2008) Y-MAC: an energy-efficient multi-channel MAC protocol for dense wireless sensor networks. In: Proceedings of the 7th IEEE international conference on information processing in sensor networks (IPSN), pp 53–63

59. Salajegheh M, Soroush H, Kalis A (2007) HyMAC: hybrid TDMA/FDMA medium access control protocol for wireless sensor networks. In: Proceedings of the 18th IEEE international symposium on personal, indoor and mobile radio communications (PIMRC)
60. So HSW, Walrand J, Mo J (2007) McMAC: a parallel rendezvous multi-channel MAC protocol. In: Proceedings of the IEEE wireless communications and networking conference (WCNC), pp 334–339
61. Wu Y, Keally M, Zhou G, Mao W (2009) Traffic-aware channel assignment in wireless sensor networks. In: Proceedings of the 4th international conference on wireless algorithms, systems, and applications (WASA), pp 479–488
62. Watteyne T, Mehta A, Pister K (2009) Reliability through frequency diversity: Why channel hopping makes sense. In: Proceedings of the 6th international symposium on performance evaluation of wireless ad hoc, sensor, and ubiquitous networks (PE-WASUN)
63. Pister K, Doherty L (2008) TSMP: time synchronized mesh protocol. In: Proceedings of the IASTED international symposium on distributed sensor networks (DSN), pp 391–398
64. Song J, Han S, Mok A, Chen D, Lucas M, Nixon M, Pratt W (2008) WirelessHART: applying wireless technology in real-time industrial process control. In: Proceedings of the 14th IEEE international real-time and embedded technology and applications symposium (RTAS), pp 377–386
65. Du P, Roussos G (2011) Adaptive channel hopping for wireless sensor networks. In: Proceedings of the IEEE international conference on selected topics in mobile and wireless networking (iCOST), pp 19–23
66. Yoon SU, Murawski R, Ekici E, Park S, Mir ZH (2010) Adaptive channel hopping for interference robust wireless sensor networks. In: Proceedings of the IEEE international conference on communications (ICC), pp 432–439
67. Chen H, Cui L, Lu S (2009) An experimental study of the multiple channels and channel switching in wireless sensor networks. In: Proceedings of the 4th international symposium on innovations and real-time applications of distributed sensor networks (IRADSN), pp 54–61
68. Won C, Youn JH, Ali H, Sharif H, Deogun J (2005) Adaptive radio channel allocation for supporting coexistence of 802.15.4 and 802.11b. In: Proceedings of the 62nd IEEE vehicular technology conference (VTC), pp 2522–2526
69. Voigt T, Österlind F, Dunkels A (2008) Improving sensor network robustness with multichannel convergecast. In: Proceedings of the 2nd ERCIM workshop on e-Mobility
70. Watteyne T, Lanzisera S, Mehta A, Pister KS (2010) Mitigating multipath fading through channel hopping in wireless sensor networks. In: Proceedings of the IEEE international conference on communications (ICC), pp 1–5
71. Kerkez B, Watteyne T, Magliocco M, Glaser S, Pister K (2009) Feasibility analysis of controller design for adaptive channel hopping. In: Proceedings of the 4th international conference on performance evaluation methodologies and tools (VALUETOOLS)
72. Xu R, Shi G, Luo J, Zhao Z, Shu Y (2011) MuZi: multi-channel zigbee networks for avoiding WiFi interference. In: Proceedings of the 4th international conference on cyber, physical and social computing (CPSCOM), pp 323–329
73. Zúñiga MA, Irzynska I, Hauer JH, Voigt T, Boano CA, Römer K (2011) Link quality ranking: getting the best out of unreliable links. In: Proceedings of the 7th IEEE international conference on distributed computing in sensor systems (DCOSS), pp 1–8
74. Radi M, Dezfouli B, Bakar KA, Lee M (2012) Multipath routing in wireless sensor networks: survey and research challenges. Sensors 12(1):650–685
75. Alizai MH, Landsiedel O, Bitsch Link JA, Götz S, Wehrle K (2009) Bursty traffic over bursty links. In: Proceedings of the 7th ACM conference on embedded networked sensor systems (SenSys), pp 71–84
76. Liu T, Cerpa A (2011) Foresee (4c): wireless link prediction using link features. In: Proceedings of the 10th IEEE international conference on information processing in sensor networks (IPSN), pp 294–305
77. Gonga A, Landsiedel O, Soldati P, Johansson M (2012) Revisiting multi-channel communication to mitigate interference and link dynamics in wireless sensor networks. In: Proceedings of the 8th IEEE international conference on distributed computing in sensor systems (DCOSS)

78. Rehmani MH, Alves T, Lohier S, Rachedi A, Poussot B (2012) Towards intelligent antenna selection in IEEE 802.15.4 wireless sensor networks. In: Proceedings of the 13th ACM international symposium on mobile ad hoc networking and computing (MobiHoc), pp 245–246
79. Giorgetti G, Cidronali A, Gupta SK, Manes G (2007) Exploiting low-cost directional antennas in 2.4 GHz IEEE 802.15.4 wireless sensor networks. In: Proceedings of the 10th European conference on wireless technologies (ECWT), pp 217–220
80. Nilsson M (2009) Directional antennas for wireless sensor networks. In: Proceedings of the 9th Scandinavian workshop on wireless adhoc network (Adhoc)
81. Öström E, Mottola L, Nilsson M, Voigt T (2010) Smart antennas made practical: the SPIDA way. In: Proceedings of the 9th ACM/IEEE international conference on information processing in sensor networks (IPSN), demo session, pp 438–439
82. Voigt T, Hewage KC, Mottola L (2013) Understanding link dynamics in wireless sensor networks with dynamically steerable directional antennas. In: Proceedings of the 10th European conference on wireless sensor networks (EWSN)
83. Yuan W, Linnartz JPM, Niemegeers IG (2010) Adaptive CCA for IEEE 802.15.4 wireless sensor networks to mitigate interference. In: Proceedings of the IEEE wireless communication and networking conference (WCNC), pp 1–5
84. Bertocco M, Gamba G, Sona A (2007) Experimental optimization of CCA thresholds in wireless sensor networks in the presence of interference. In: Proceedings of the IEEE Europe the workshop on electromagnetic compatibility (EMC)
85. Woehrle M (2010) Testing of wireless sensor networks. Ph.D. thesis, Eidgenössische Technische Hochschule (ETH), Zürich, Switzerland
86. Burchfield R, Nourbakhsh E, Dix J, Sahu K, Venkatesan S, Prakash R (2009) RF in the jungle: effect of environment assumptions on wireless experiment repeatability. In: Proceedings of the IEEE international conference on communications (ICC), pp 4993–4998
87. Kamthe A, nán MACP, Cerpa AE (2009) M&M: multi-level Markov model for wireless link simulations. In: Proceedings of the 7th ACM conference on embedded networked sensor systems (SenSys), pp 57–70
88. Lee H, Cerpa A, Levis P (2007) Improving wireless simulation through noise modeling. In: Proceedings of the 6th international conference on information processing in sensor networks (IPSN '07). ACM, New York, pp 21–30
89. Zhou G, He T, Krishnamurthy S, Stankovic J (2006) Models and solutions for radio irregularity in wireless sensor networks. ACM Trans Sens Netw (TOSN) 2:221–262
90. Garg K, Förster A, Puccinelli D, Giordano S (2011) Towards realistic and credible wireless sensor network evaluation. In: Proceedings of the 3rd international conference on ad hoc networks (ADHOCNETS), pp 49–64
91. Boers NM, Nikolaidis I, Gburzynski P (2012) Impulsive interference avoidance in dense wireless sensor networks. In: Proceedings of the 11th international conference on ad-hoc networks and wireless (AdHocNow)
92. Boano CA, Römer K, Österlind F, Voigt T (2011) Realistic simulation of radio interference in COOJA. In: Adjunct proceedings of the 8th European conference on wireless sensor networks (EWSN), demo session, pp 36–37
93. Österlind F (2011) Improving low-power wireless protocols with timing-accurate simulation. Ph.D. thesis, Uppsala University, Uppsala, Sweden
94. Nethi S, Nieminen J, Jäntti R (2011) Exploitation of multi-channel communications in industrial wireless sensor applications: avoiding interference and enabling coexistence. In: Proceedings of the IEEE wireless communications and networking conference (WCNC), pp 345–350
95. Shin SY, Kang JS, Park HS (2009) Packet error rate analysis of zigbee under interferences of multiple bluetooth piconets. In: Proceedings of the 69th IEEE vehicular technology conference (VTC)
96. Iyer V, Woehrle M, Langendoen K (2010) Chamaeleon: exploiting multiple channels to mitigate interference. In: Proceedings of the 7th international conference on networked sensing systems (INSS), pp 65–68

97. Bertocco M, Gamba G, Sona A, Tramarin F (2008) Investigating wireless networks coexistence issues through an interference aware simulator. In: Proceedings of the 13th IEEE international conference on emerging technologies and factory automation (ETFA), pp 1153–1156
98. Dominicis CMD, Ferrari P, Flammini A, Sisinni E, Bertocco M, Giorgi G, Narduzzi C, Tramarin F (2009) Investigating WirelessHART coexistence issues through a specifically designed simulator. In: Proceedings of the IEEE international instrumentation and measurement technology conference (I2MTC), pp 1085–1090
99. Zhao Z, Yang GH, Liu Q, Li V, Cui L (2010) Easitest: a multi-radio testbed for heterogeneous wireless sensor networks. In: Proceedings of the IET international conference on wireless sensor network (IET-WSN), pp 104–108
100. Sanchez A, Moerman I, Bouckaert S, Willkomm D, Hauer JH, Michailow N, Fettweis G, Dasilva L, Tallon J, Pollin S (2011) Testbed federation: an approach for experimentation-driven research in cognitive radios and cognitive networking. In: Proceedings of the of the 20th future network and mobile summit
101. Xu W, Trappe W, Zhang Y (2008) Defending wireless sensor networks from radio interference through channel adaptation. ACM Trans Sens Netw (TOSN) 4:1–34
102. Zhou G, Lu J, Wan CY, Yarvis MD, Stankovic JA (2008) BodyQoS: adaptive and radio-agnostic QoS for body sensor networks. In: Proceedings of the 27th IEEE international conference on computer communications (INFOCOM), pp 565–573
103. Sakamuri D (2008) NetEye: a wireless sensor network testbed. Master's thesis, Wayne State University, Detroit, Michigan
104. Khaleel H, Pastrone C, Penna F, Spirito M, Garello R (2009) Impact of Wi-Fi traffic on the IEEE 802.15.4 channels occupation in indoor environments. In: Proceedings of the 11th international conference on electromagnetics in advanced applications (ICEAA)
105. Gnawali O, Guibas L, Levis P (2010) Case for evaluating sensor network protocols concurrently. In: Proceedings of the 5th ACM international workshop on wireless network testbeds, experimental evaluation and characterization (WiNTECH), pp 47–54
106. Ortiz J, Culler D (2008) Exploring diversity: evaluating the cost of frequency diversity in communication and routing. In: Proceedings of the 6th ACM international conference on embedded networked sensor systems (SenSys), poster session, pp 411–412
107. Ortiz J, Culler D (2010) Multichannel reliability assessment in real world WSNs. In: Proceedings of the 9th IEEE international conference on information processing in sensor networks (IPSN), pp 162–173
108. Stabellini L (2010) Energy-aware channel selection for cognitive wireless sensor networks. In: Proceedings of the 7th international symposium on wireless communication systems (ISWCS), pp 892–896
109. Gnawali O, Fonseca R, Jamieson K, Moss D, Levis P (2009) Collection tree protocol. In: Proceedings of the 7th ACM conference on embedded networked sensor systems (SenSys '09). ACM, New York, pp 90–100
110. Stabellini L, Parhizkar MM (2010) Experimental comparison of frequency hopping techniques for 802.15.4-based sensor networks. In: Proceedings of the 4th international conference on mobile ubiquitous computing, systems, services and technologies (UBICOMM), pp 110–116
111. Zhou G, Lu L, Krishnamurthy S, Keally M, Ren Z (2009) SAS: self-adaptive spectrum management for wireless sensor networks. In: Proceedings of the 18th internatonal conference on computer communications and networks (ICCCN), pp 1–6
112. Sha M, Hackmann G, Lu C (2011) Multi-channel reliability and spectrum usage in real homes: empirical studies for home-area sensor networks. In: Proceedings of the 19th IEEE international workshop on quality of service (IWQoS), pp 39:1–39:9
113. Dutta P, Dawson-Haggerty S, Chen Y, Liang CJM, Terzis A (2010) Design and evaluation of a versatile and efficient receiver-initiated link layer for low-power wireless. In: Proceedings of the 8th ACM international conference on embedded networked sensor systems (SenSys), pp 1–14

114. Moeller S, Sridharan A, Krishnamachari B, Gnawali O (2010) Routing without routes: the backpressure collection protocol. In: Proceedings of the 9th international conference on information processing in sensor networks (IPSN), pp 279–290
115. Lau SY, Lin TH, Huang TY, Ng IH, Huang P (2009) A measurement study of zigbee-based indoor localization systems under RF interference. In: Proceedings of the 4th ACM international workshop on experimental evaluation and characterization (WINTECH), pp 35–42
116. Kang MS, Chong JW, Hyun H, Kim SM, Jung BH, Sung DK (2007) Adaptive interference-aware multi-channel clustering algorithm in a zigbee network in the presence of WLAN interference. In: Proceedings of the 2nd international symposium on wireless pervasive computing (ISWPC)
117. Rohde J, Toftegaard TS (2011) Mitigating the impact of high interference levels on energy consumption in wireless sensor networks. In: Proceedings of the 2nd international conference on wireless communication, vehicular technology, information theory and aerospace and electronic systems technology (Wireless VITAE), pp 1–5
118. Ahmed N, Kanhere SS, Jha S (2010) Mitigating the effect of interference in wireless sensor networks. In: Proceedings of the 35th IEEE international conference on local computer networks (LCN), pp 160–167
119. Ahmed N, Kanhere SS, Jha S (2010) Experimental evaluation of multi-hop routing protocols for wireless sensor networks. In: Proceedings of the 9th ACM/IEEE international conference on information processing in sensor networks (IPSN), pp 416–417
120. Ansari J, Ang T, Mähönen P (2011) WiSpot: fast and reliable detection of Wi-Fi networks using IEEE 802.15.4 radios. In: Proceedings of the 9th ACM international workshop on mobility management and wireless access (MobiWac), pp 35–44
121. Hossian MA, Mahmood A, Jäntti R (2009) Channel ranking algorithms for cognitive coexistence of IEEE 802.15.4. In: Proceedings of the 20th IEEE international symposium on personal, indoor and mobile radio communications (PIMRC), pp 112–116
122. Huang GJ, Xing G, Zhou G, Zhou R (2010) Beyond co-existence: exploiting WiFi white space for zigbee performance assurance. In: Proceedings of the 18th IEEE international conference on network protocols (ICNP), pp 305–314
123. Shuaib K, Boulmalf M, Sallabi F, Lakas A (2006) Co-existence of zigbee and WLAN: a performance study. In: Proceedings of the IEEE international wireless telecommunications symposium (WTS)
124. Jeong Y, Kim J, Han SJ (2011) Interference mitigation in wireless sensor networks using dual heterogeneous radios. Wirel Netw 17(7):1699–1713
125. Arkoulis S, Spanos DE, Barbounakis S, Zafeiropoulos A, Mitrou N (2010) Cognitive radio-aided wireless sensor networks for emergency response. Meas Sci Technol 21(12):124002
126. Ansari J, Ang T, Mähönen P (2010) Spectrum agile medium access control protocol for wireless sensor networks. In: Proceedings of the 7th IEEE international conference on sensor mesh and ad hoc communications and networks (SECON), pp 1–9
127. Qin Y, He Z, Voigt T (2011) Towards accurate and agile link quality estimation in wireless sensor networks. In: Proceedings of the 10th IFIP annual mediterranean ad-hoc networking workshop (Med-Hoc-Net), pp 179–185
128. Boano CA, He Z, Li Y, Voigt T, Zúñiga M, Willig A (2009) Controllable radio interference for experimental and testing purposes in wireless sensor networks. In: Proceedings of the 4th international workshop on practical issues in building sensor network applications (SenseApp), pp 865–872
129. Duquennoy S, Österlind F, Dunkels A (2011) Lossy links, low power, high throughput. In: Proceedings of the 9th ACM conference on embedded networked sensor systems (SenSys), pp 12–25
130. Fotouhi H, Zúñiga MA, Alves M, Koubâa A, Marrón PJ (2012) Smart-HOP: a reliable hand-off mechanism for mobile wireless sensor networks. In: Proceedings of the 9th European conference on wireless sensor networks (EWSN), pp 131–146
131. Österlind F, Mottola L, Voigt T, Tsiftes N, Dunkels A (2012) Strawman: resolving collisions in bursty low-power wireless networks. In: Proceedings of the 11th international conference on information processing in sensor networks (IPSN), pp 161–172

132. Puccinelli D, Gnawali O, Yoon S, Santini S, Colesanti U, Giordano S, Guibas L (2011) The impact of network topology on collection performance. In: Proceedings of the 8th European conference on wireless sensor networks, EWSN'11. Springer, Berlin, pp 17–32
133. Akan OB, Karli OB, Ergul O (2009) Cognitive radio sensor networks. IEEE Network: The Magazine of Global Internetworking 23(4):34–40

Chapter 3
Overview of Link Quality Estimation

Abstract Low-power links exhibit a complex and dynamic behavior, e.g., frequent quality fluctuations, highly asymmetric connectivity, large transitional region. They are more unreliable than traditional wireless links. This is due to the use of low-power and low-cost radio transceivers. These facts raised the need for link quality estimation as a fundamental building block for higher-layer protocols. This chapter presents the fundamental concepts related to link quality estimation in low-power wireless networks, such as the definition of the link quality estimation process, and the requirements for efficient Link Quality Estimators (LQEs). This chapter gives also a comprehensive survey of existing LQEs.

3.1 Introduction

Link quality estimation enables network protocols to mitigate and to overcome link unreliability. For instance, link quality estimation is instrumental for routing protocols to maintain correct network operation [1–8]. Delivering data over high quality links (i) improves the network delivery by limiting packet loss and (ii) maximizes its lifetime by minimizing the number of retransmissions, and avoiding route reselection triggered by link failures. Link quality estimation also plays a crucial role for topology-control mechanisms to maintain the stability of the topology [9–11]. High quality links are long-lived, therefore, efficient topology control mechanisms rely on the selection of high quality links in order to maintain robust network connectivity for long periods, thus avoiding unwanted transient topology breakdowns.

This chapter outlines the fundamental concepts related to link quality estimation in low-power wireless networks. First, we present link quality estimation as a stepwise process. Then, we discuss the requirements for the design of efficient Link Quality Estimators (LQEs). Finally, we give a comprehensive survey of existing LQEs. In this survey, LQEs are classified in two categories: hardware-based LQEs and software-based LQEs.

N. Baccour et al., *Radio Link Quality Estimation in Low-Power Wireless Networks*,
SpringerBriefs in Electrical and Computer Engineering,
DOI: 10.1007/978-3-319-00774-8_3,

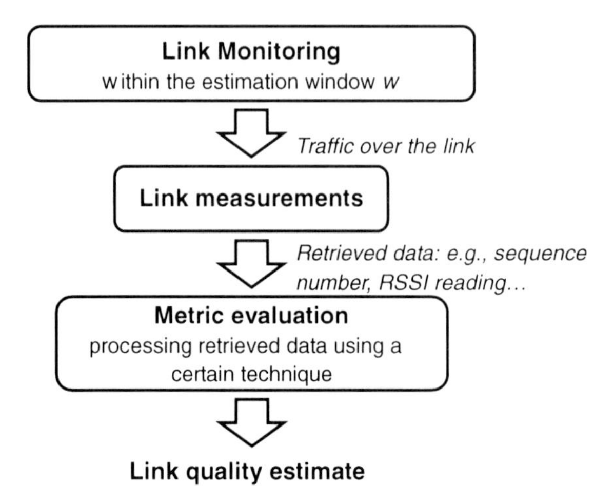

Fig. 3.1 Steps for link quality estimation

3.2 The Link Quality Estimation Process

Basically, link quality estimation consists of evaluating a *metric*—a mathematical expression, within an estimation window w (e.g., at each w seconds, or based on w received/sent packets). We refer to this metric as Link Quality Estimator (LQE). The LQE evaluation requires *link measurements*. For example, to evaluate the PRR estimator, link measurements consist of extracting the sequence number from each received packet. *Link monitoring* defines a strategy to have traffic over the link allowing for link measurements. Hence, the link quality estimation process involves three steps: link monitoring, link measurements, and metric evaluation. These steps are described next and illustrated in Fig. 3.1.

3.2.1 Link Monitoring

There are three kinds of link monitoring: (i) active link monitoring, (ii) passive link monitoring, and (iii) hybrid link monitoring. Note that not only link quality estimation relies on link monitoring, but also other mechanisms, such as routing and topology control [3].

Active link monitoring: In active link monitoring, a node monitors the links to its neighbors by sending probe packets. Probe packets can be sent either by broadcast such as in [12], or by unicast such as in [13]. Broadcast probe packets involve no link-level acknowledgments or retransmissions, in contrast to unicast probe packets. Probe packets are generally sent at a certain rate, which yields a tradeoff between energy-efficiency (low rates) and accuracy (high rates). An adaptive beaconing rate, such as the one proposed in [3] might provide a good balance for this tradeoff.

Broadcast-based active link monitoring is simple to implement and incurs a small overhead compared to unicast-based [13]. For that reason, many network protocols and mechanisms rely on it. On the other hand, unicast-based active link monitoring enables more *accurate* link measurements because of its resemblance to actual data transmission over the link [14]. However, it is still considered to be a costly mechanism for low-power wireless networks due to the communication overhead.

Passive link monitoring: Unlike active link monitoring, passive link monitoring exploits existing traffic without incurring additional communication overhead. In fact, a node listens to transmitted packets, even if these packets are not addressed to it (overhearing) [15, 16]. It can also listen to acknowledgments of messages sent by different neighbors [1, 4].

Passive link monitoring has been widely used in low-power wireless networks due to its energy-efficiency compared to active link monitoring [4, 15–20]. However, passive monitoring incurs the overhead of probing idle links [13]. Lal et al. [15] found that overhearing involves a significant energy expenditure. In addition, when the network operates at a low data rate or with unbalanced traffic, passive link monitoring may lead to the lack of up-to-date link measurements. Consequently, it leads to inaccurate link quality estimation.

Hybrid link monitoring: The use of a hybrid mechanism combining both active and passive monitoring may yield an efficient balance between up-to-date link measurements and energy-efficiency [13]. For instance, in [3], the authors introduced a hybrid link monitoring mechanism for performing both link quality estimation and routing advertisements. Active link monitoring consists in broadcasting beacons with a non-fixed rate. Rather, a specific algorithm is used to adaptively tune the beaconing rate: initially, the beaconing rate is high and decreases exponentially until it reaches a certain threshold. When the routing layer signals some problems such as loop detection, the beaconing rate resets to its initial value. Active link monitoring is coupled with passive link monitoring, which consists in hearing received acknowledgments from neighbours (that represent next hops).

Finally, it has been shown in several recent studies that link quality estimation where link monitoring is based on data traffic is much more accurate than that having link monitoring based on beacon traffic [3, 14, 21, 22]. The reason is that there are several differences between unicast and broadcast link properties [22]. It is thereby difficult to precisely estimate unicast link properties via those of broadcast.

3.2.2 Link Measurements

Link measurements are performed by retrieving useful information (i) from received packets/acknowlegments or (ii) from sent packets. Data retrieved from received packets/acknowledgments, such as sequence numbers, time stamp, RSSI, and LQI, is used to compute *receiver-side* link quality estimators. On the other hand, data retrieved from sent packets, e.g., sequence numbers, time stamp and packet retransmission count, allows for the computation of *sender-side* link quality estimators.

3.2.3 Metric Evaluation

Based on link measurements, a metric is evaluated to produce an estimation of the link quality. Generally, this metric is designed according to a certain *estimation technique*, which can be a simple average or a more sophisticated technique such as filtering, learning, regression, Fuzzy Logic, etc. For example, Woo et al. [2] introduced the WMEWMA estimator, which uses the EWMA filter as main estimation technique: based on link measurements, the PRR is computed and then smoothed to the previously computed PRR using EWMA filter. More examples are given in Sect. 3.5 and Table 3.1.

3.3 Design Requirements

The design of efficient LQEs has several requirements, which are described next.

3.3.1 Energy Efficiency

As energy may be a major concern in low-power wireless networks, LQEs should involve low computation and communication overhead. Consequently, some complex estimation techniques such as learning might be not appropriate in low-power wireless networks. Moreover, LQEs should also involve low communication overhead. Typically, an active monitoring with high beaconing rate should be avoided as it is energy consuming.

3.3.2 Accuracy

It refers to the ability of the LQE to correctly characterize the link state, i.e., to capture the real behavior of the link. The accuracy of link quality estimation greatly impacts the effectiveness of network protocols. In traditional estimation theory, an estimated process is typically compared to a real known process using a certain statistical figure (e.g., least mean square error or regression analysis). However, such comparison is not possible in link quality estimation, since: (i) there is no metric that is widely considered as the "real" figure to measure link quality; and (ii) link quality is represented by quantities of different nature: some estimators are based on the computation of packet reception ratio, some are based on packet retransmission count, and some are hybrid of these, as described in 3.5. Nevertheless, the accuracy of LQEs can be assessed indirectly, i.e., resorting to a metric that subsume the effect of link quality estimation. For instance, in [23], the authors

Table 3.1 Comparison and classification of LQEs

			Technique	Link asymmetry support	Monitoring	Location
Hardware-based	RSSI, LQI and SNR		Read from hardware and can be averaged	No	Passive or Active	Receiver
Software-based	**PRR-based**	*PRR*	Average	No	Passive or Active	Receiver
		WMEWMA [16]	Filtering	No	Passive	Receiver
		KLE [28]	Filtering	No	–	Receiver
	RNP-based	*RNP* [17]	Average	No	Passive	Sender
		LI [15]	Probability	No	Passive	Receiver
		ETX [12]	Average	Yes	Active	Receiver
		four-bit [23]	Filtering	Yes	Active and Passive	Sender and receiver
		L-NT and L-ETX [14]	Filtering	No	–	Sender
	Score-based	WRE [18]	Regression	No	Passive	Receiver
		MetricMap [20]	Training and classification	No	Passive	Receiver
		F-LQE [27]	Fuzzy logic	Yes	Passive	Receiver
		CSI [29]	Weighted sum	No	Active	Receiver

studied the impact of their four-bit LQE on the performance of CTP (Collection Tree Protocol), a hierarchical routing protocol [3]. They found that four-bit leads to better end-to-end packet delivery ratio, compared with the original version of CTP. Hence, four-bit might be more accurate as it can correctly select routes composed of high quality links. On the other hand, the authors in [24] analyzed the accuracy (referred as reliability) of LQEs by analyzing their statistical properties, namely their temporal behavior and the distribution of link quality estimates.

3.3.3 Reactivity

It refers to the ability to quickly react to persistent changes in link quality [25]. For example, a reactive LQE enables routing protocols and topology control mechanisms to quickly adapt to changes in the underlying connectivity. Reactivity depends on two factors: the estimation window w and the link monitoring scheme. Low w and active monitoring with high beaconing rate can lead to reactive LQE. Though, it is important to note that some LQEs are naturally more reactive than others regardless of the w value or the link monitoring schema. In fact, LQEs that are computed at the sender-side were shown to be more reactive than those computed at the receiver-side [24]. More details are given in 3.5.

3.3.4 Stability

It refers to the ability to tolerate transient (short-term) variations in link quality. For instance, routing protocols do not have to recompute information when a link quality shows transient degradation, because rerouting is a very energy and time consuming operation. Lin et al. [26] argued that stability is met through long-term link quality estimation. Long-term link quality estimation was performed by the means of the EWMA filter with a large smoothing factor ($\alpha = 0.9$). Hence, they introduced *Competence* metric that applies the EWMA filter to a binary function indicating whether the current measured link quality is within a desired range. Stability of LQEs can be assessed by the coefficient of variation of link quality estimates, which is computed as the ratio of the standard deviation to the mean [16]. It can also be assessed by studying the impact of the LQE on routing, typically a stable LQE leads to stable topology, e.g., few parent changes in the case of hierarchical routing [27].

As a matter of fact, reactivity and stability are at odds. For instance, consider using PRR as LQE, if we compute the PRR frequently (small w), we obtain a reactive LQE as it captures link dynamics at a fine grain. However, this reliability will be at the cost of stability because the PRR will consider some transient link quality fluctuation that might be ignored. Thus, a good LQE is the one that provides a good tradeoff between reactivity and stability. Lin et al. [26] suggest combining their long-term metric Competence, considered as a stable but not reactive LQE, with a short-term

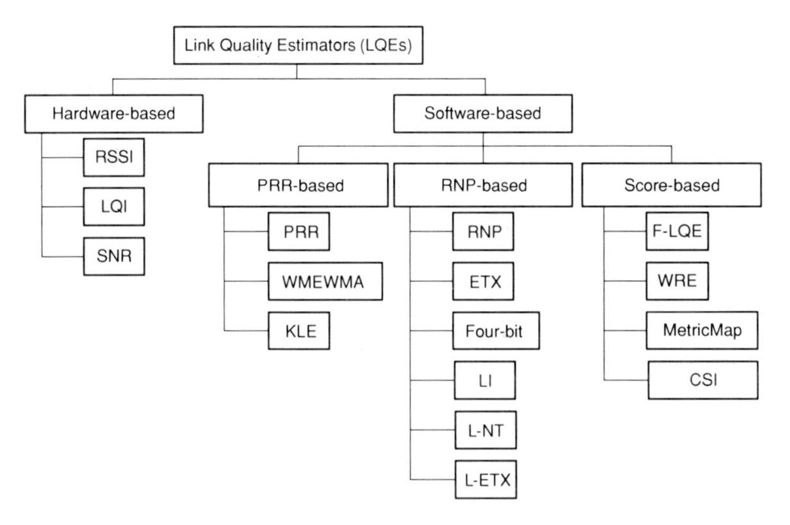

Fig. 3.2 Taxonomy of LQEs

metric such as ETX, considered as a reactive but unstable LQE, to obtain a good tradeoff. They introduced routing schemes based on this principle. For example, in a tree-based routing scheme, a node selects a potential parent as the neighbour having the best Competent link, among all neighbours having low route cost, where route cost is computed based on ETX. The authors argued that such routing scheme selects links that are good in both the short and the long term, and leads to stable network performance. On the other hand, Woo et al. [2] argued that using EWMA filter with convenient smoothing factor would strike balance between reactivity and stability.

Several efforts were carried out for the design of efficient LQEs. Next, we survey, and discuss the most relevant LQEs that are suitable for low-power wireless networks. These LQEs can be classified in two categories: hardware-based and software-based, as illustrated in Fig. 3.2. Table 3.1 presents a comparison of LQEs in low-power wireless networks.

3.4 Hardware-Based LQEs

Hardware-based LQEs are directly read from the radio transceiver (e.g., the CC2420), i.e., they do not require any additional computation. Three LQEs belong to the family of hardware-based LQEs: LQI, RSSI, and SNR.[1] The adequacy of hardware-based LQEs in characterizing links was subject of several research work. We have summarized the literature related to this issue in the following observations:

[1] Some radio transceivers do not provide LQI.

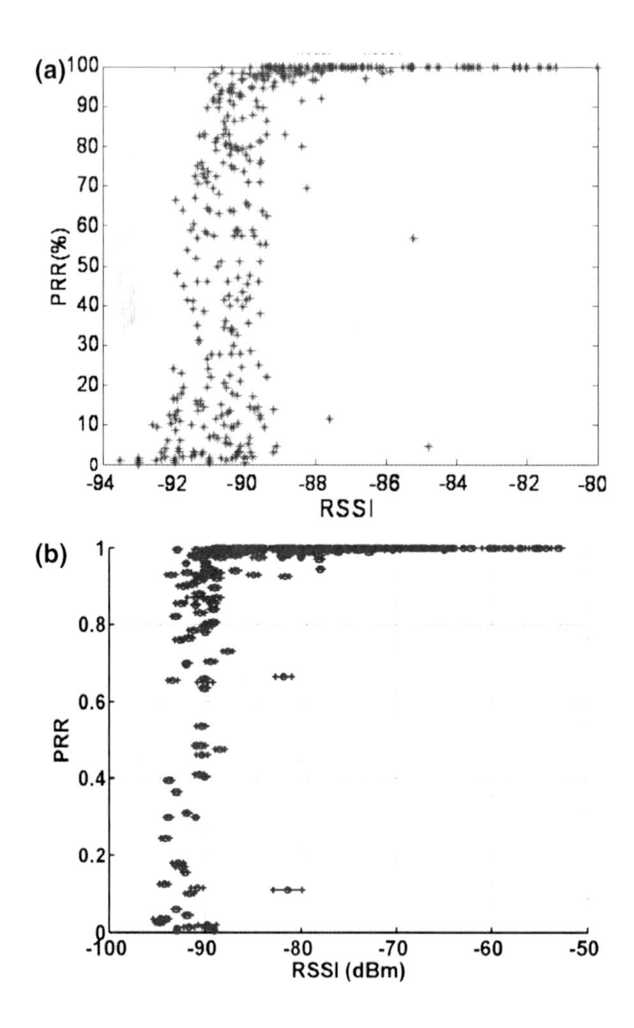

Fig. 3.3 PRR versus RSSI curve. **a** Outdoor environment, using TelosB sensor motes (using the RadiaLE testbed [24]). **b** Indoor environment, using MicaZ sensor motes [30]

Observation 1: RSSI can provide a quick and accurate estimate of whether a link is of very good quality (connected region). This observation was justified by the following: First, empirical studies such as [30] proved the existence of a RSSI value (-87 dBm [30]) above which the PRR is consistently high (99 % [30]), i.e., belong to the connected region. Below this threshold, a shift in the RSSI as small as 2 dBm can change a good link to a bad one and vice versa, which means that the link is in the transitional or disconnected region [31]. This observation is illustrated in Fig. 3.3a and b. Second, RSSI was shown very stable (standard deviation less than 1 dBm) over a short time span (2 s), thereby a single RSSI reading (over a packet reception) is sufficient to determine if the link is in the transitional region or not [31].

Observation 2: LQI can determine whether the link is of very good quality or not. However, it is not a good indicator of intermediate quality links due to its high variance, unless it is averaged over a certain number of readings. Srinivasan et al. [31] argued that when the LQI is very high (near 110) the link is of perfect quality (near 100 % of PRR). Further, in this situation LQI has low variance so that a single LQI reading would be sufficient to decide if the link is of perfect quality or not. On the other hand, for other LQI values, corresponding to intermediate quality links, the variance of LQI becomes significant and a single LQI reading is not sufficient for accurate link quality estimation. Srinivasan and Levis [32] showed that LQI should be averaged over a large packet window (about 40 up to 120 packets) to provide accurate link quality estimation, but this will be at the cost of agility and responsiveness to link quality changes. The LQI high variance is due to the fact that LQI is a statistical value [30].

Bringing observations 1 and 2 together, it might be reasonable to use a single RSSI or LQI reading to decide if the link is of high quality or not. Such decision is based on RSSI and LQI thresholds, beyond which a link can maintain high quality, e.g., a PRR of at least 95 % [33]. Importantly, these thresholds depend on the environment characteristics. For example, Lin et al. [33] found that RSSI threshold is around -90 dBm on a grass field, -91 dBm on a parking lot, and -89 dBm in a corridor. For LQI and RSSI values below these thresholds, neither of these metrics can be used to differentiate links clearly. Nevertheless, an average LQI, with the convenient averaging window, allows a more accurate classification of intermediate links [32]. On the other hand, Mottola et al. [34] claimed that RSSI should not be used to classify intermediate links.

Observation 3: The variance of LQI can be exploited for link quality estimation. Empirical studies [32] pointed out that links of intermediate and bad quality have high LQI variance, therefore the LQI needs to be averaged over many samples to give meaningful results. Boano et al. [35] proposed the use of the variance of LQI to distinguish between good links, having very low LQI variance and bad links, having very high LQI variance using as few as 10 samples. However, in that work, the authors did not provide a mapping function or a mathematical expression that exploit the variance of LQI to provide a link quality estimate.

Observation 4: LQI is a better indicator of the PRR than RSSI. In [32, 36, 37], it was argued that average LQI shows stronger correlation with PRR, compared to average RSSI. Hence, LQI is a better indicator of PRR than RSSI. On the other hand, in [32] and [36], the authors claimed that RSSI has the advantage of being more stable than LQI (i.e., it shows lower variance), except for multi-path affected links. In fact, which of LQI and RSSI is better for link quality estimation is an unanswered question, reflected by several contradicting statements and results.

Observation 5: SNR is a good indicator and even predictor of the PRR but it is not accurate, especially for intermediate links. Theoretically, for a given modulation schema, the SNR leads to an expected bit error rate, which can be extrapolated to packet error rate and then to the PRR [38]. Hence, an analytical expression that gives the PRR as a function of SNR can be derived [38]. Srinivasan et al. [31] justified the observed link characteristics (e.g., link temporal variation and link asymmetry) with SNR behavior. Particularly, they assume that changes in PRR must be due to changes in SNR. However, other studies [19, 28, 39] showed that the theoretical relationship between SNR and PRR reveals many difficulties. These difficulties arise from the fact that mapping between SNR and PRR depends on the actual sensor hardware and environmental effects such as temperature [28]. As a result, these studies concluded that SNR cannot be used as a standalone estimator, but it may help to enhance the accuracy of the PRR estimation. Further, Lal et al. [15] recommended not to use SNR as link quality estimator, when links are inside the transitional region.

Observation 6: SNR is a better link quality estimator than RSSI. The RSSI is the sum of the pure received signal and the noise floor at the receiver. On the other hand, the SNR describes how strong the pure received signal is in comparison with the receiver noise floor. As the noise floor at different nodes can be different, the SNR metric should be better than RSSI [31].

Hardware-based LQEs share some limitations: first, these metrics are only measured for successfully received packets; thus, when a radio link suffers from excessive packet losses, they may overestimate the link quality by not considering the information of lost packets. Second, despite the fact that hardware metrics provide a fast and inexpensive way to classify links as either good or bad, they are incapable of providing a fine grain estimation of link quality [23, 40].

The above limitations of hardware-based LQEs do not mean that this category of LQEs is not useful. In fact, each of these LQEs provides a particular information on the link state, but none of them is able to provide a holistic characterization of the link quality. Currently, there is a growing awareness that the combination of hardware metrics with software metrics can improve the accuracy of the link quality estimation [23, 40–43]. For example, Fonseca et al. [23] use LQI as a hardware metric to quickly decide whether the link is of good quality. If it is the case, the node is included in the *neighbor table* together with the link quality, assessed using Four-bit as a software metric. Gomez et al. [40] confirm that LQI can accurately identify high quality links, but it fails to accurately classify intermediate links due to its high variance. They exploited this observation to design LETX (LQI-based ETX), a link estimator that is dedicated for routing. The authors first build a piecewise linear model of the PRR as a function of average LQI. This model allows to estimate the PRR given one LQI sample. LETX is then computed as the inverse of the estimated PRR. LETX is used to identify high quality links in route selection process. Rondinone et al. [42] also suggest combining hardware and software metrics though a multiplicative metric between PRR and RSSI, and Boano et al. [43] propose a fast estimator suitable for mobile environments by combining geometrically PRR, SNR, and LQI.

The discussion above has been constrained to the hardware-based LQEs that are provided by current low power radio transceivers such as the CC2420 radio. This radio has been used in almost all studies regarding LQE that have been conducted. In order to be independent of the current transceiver technology, Qin et al. have used Software-Defined Radios (SDR) to explore hardware-based LQEs [44, 45]. Among the estimators they propose, a particular interesting one is the "Spectrum Factor" (SF).

The SF metric for the 5 MHz-wide IEEE 802.15.4 signals takes the ratio between the energy of the main lobe and that of the sum of the two side lobes [45]. The SF does not rely on successful synchronization or demodulation and hence can be computed even on unsuccessfully demodulated packets. This makes SF in particular useful to estimate very weak links.

The main advantage of using SDR for link quality estimation is that, as mentioned above, one is not constrained to existing radio hardware and its limitations. For example, the LQI in the CC2420 radio is only computed based on a limited number of symbols. Furthermore, little processing power is often assumed but it is not unlikely that future smart objects might incorporate more signal processing capabilities. Therefore, SDR is a very useful tool to explore new metrics that are not implemented in today's radios. LQE research based on SDR could, in the best case, even provide radio manufacturers with new or improved metrics for future generation of low power radios. The downside of the SDR approach is that the hardware is costly and the learning curve for SDR is quite steep.

3.5 Software-Based LQEs

In contrast to hardware-based LQEs, which are directly obtained from the radio transceiver, software-based LQEs are derived through packet-statistics collection (e.g., packet sequence number). Software-based LQEs can be classified into three categories, as illustrated in Fig. 3.2: (i) PRR-based: either count or approximate the PRR, (ii) RNP-based: either count or approximate the RNP (Required Number of Packet retransmissions), and (iii) Score-based: provide a score identifying the link quality. Table 3.1 summarizes the main characteristics of these LQEs.

3.5.1 PRR-Based

PRR can be computed as the ratio of the number of successfully received packets to the number of transmitted packets and can be computed at the receiver side, for each window of w received packets, as:

$$PRR(w) = \frac{Number\ of\ received\ packets}{Number\ of\ sent\ packets} \tag{3.1}$$

PRR is simple to measure and was widely used in routing protocols [1, 12]. Further, it was often used as an unbiased metric to evaluate the accuracy of hardware-based estimators. In fact, a hardware-based estimator that correlates with PRR is considered as a good metric.

Discussion: The efficiency of PRR depends on the adjustment of the time window size. Cerpa et al. [11] showed that for links with very high or very low PRRs, accurate link quality estimation can be achieved within narrow time windows. On the other hand, links with medium PRRs need much larger time windows to converge to an accurate link quality estimation.

The objective of LQEs that *approximate* the PRR is to provide more efficient link quality estimates than the PRR. In the following, we review the most relevant LQEs in this category.

The Window Mean with Exponentially Weighted Moving Average (WMEWMA) [16] is a receiver-side LQE based on passive monitoring. It applies EWMA filter on PRR to smooth it, thus providing a metric that resists to transient fluctuation of PRRs, yet is responsive to major link quality changes. WMEWMA is then given by the following:

$$WMEWMA(\alpha, w) = \alpha \times WMEWMA + (1 - \alpha) \times PRR \tag{3.2}$$

where $\alpha \,\varepsilon\, [0..1]$ controls the smoothness. This factor enables to give more importance, to the current PRR value (with $\alpha < 0.5$) or to the last computed WMEWMA value (with $\alpha > 0.5$).

Discussion: To assess the performance of WMEWMA, Woo and Culler [16] introduced a set of LQEs that approximate the PRR using filtering techniques other than EWMA. Then, they compared WMEWMA to these filter-based LQEs, in terms of (i) reactivity assessed by the settling time and the crossing time, (ii) accuracy evaluated by the mean square error, (iii) stability assessed by the coefficient of variation, and (iv) efficiency assessed by the memory footprint and computation complexity. WMEWMA was found to outperform the other filter-based LQEs. The work by Woo and Culler [16] laid the foundation for subsequent work on filter-based LQE, although their solution required a more thorough assessment, e.g., based on real-world data traces instead of synthetic ones (i.e., generated analytically).

The Kalman filter based link quality estimator (KLE) [28] was proposed to overcome the poor reactivity of average-based LQEs, including PRR. In fact, the objective of KLE is to provide a link quality estimate based on a single received packet rather than waiting for the reception of a certain number of packets within the estimation window and then compute the average. Upon packet reception, RSS (Received Signal Strength) is extracted and injected to a Kalman filter, which produces an estimation of the RSS. Then, an approximation of the SNR is gathered by subtracting the noise floor estimate from the estimated RSS. Using a pre-calibrated PRR-SNR curve at the receiver, the approximated SNR is mapped to an approximated PRR, which represents the KLE link quality estimate.

Discussion: Through experiments using a low-power wireless networks platform of two nodes (a sender and a receiver), Senel et al. [28] proved that KER is able to detect link quality changes faster (i.e., it is more reactive) than PRR. However, the accuracy of KER was not examined. This accuracy is typically related to the accuracy of the PRR-SNR curve, which was considered as constant over time. According to empirical observations on low-power links, this curve varies over time (in dynamic environments) and also from one node to another. Further, it seems that the positive results found by Senel et al. [28] related to the reactivity of KER are due to the steady environment in the experimental evaluation, so that the PRR-SNR curve is constant over time.

3.5.2 RNP-Based

The Required Number of Packet transmissions (RNP) [17] is a sender-side estimator that counts the average number of packet transmissions/re-transmissions, required before successful reception. Based on passive monitoring, this metric is evaluated at the sender side for each w transmitted and re-transmitted packets, as follows:

$$RNP(w) = \frac{\textit{Number of transmitted and retransmitted packets}}{\textit{number of successfully received packets}} - 1 \tag{3.3}$$

Note that the number of successfully received packets is determined by the sender as the number of acknowledged packets. RNP assumes an ARQ (Automatic Repeat Request) protocol [46] at the link-layer level, i.e., a node will repeat the transmission of a packet until it is correctly received. Note that a similar metric to the RNP is the Acknowledgment Reception Ratio (ARR). It is computed as the ratio of the number of acknowledged packets to the total number of transmitted packets during a predefined time window.

Discussion: Cerpa et al. [17] argued that RNP is better than PRR for characterizing the link quality. In fact, as opposed to RNP, PRR provides a coarse-grain estimation of link quality since it does not take into account the underlying distribution of losses. However, RNP has the disadvantage of being very unstable and can not reliably estimate the link packet delivery, mainly due to link asymmetry [27].

In the following, we review the most relevant LQEs that approximate the RNP using several techniques.

The Expected Transmission Count (ETX) [12] is a receiver-initiated estimator that approximates the packet retransmissions count. It uses active monitoring, which means that each node explicitly broadcasts probe packets to collect statistical information. ETX takes into account link asymmetry by estimating the uplink quality from the sender to the receiver, denoted as $PRR_{forward}$, as well as the downlink quality from the receiver to the sender, denoted as $PRR_{backward}$. The combination of both PRR estimates provides an estimation of the bidirectional link quality, expressed as:

$$ETX(w) = \frac{1}{PRR_{forward} \times PRR_{backward}} \tag{3.4}$$

Note that $PRR_{forward}$ is simply the PRR of the uplink determined at the receiver, for each w received probe packets, while $PRR_{backward}$ is the PRR of the downlink computed at the sender and sent to the receiver in the probe packet.

Discussion: It was shown that routing protocols based on the ETX metric provide high-throughput routes on multi-hop wireless networks, since ETX minimizes the expected total number of packet transmissions required to successfully deliver a packet to the destination [12]. In [20], the authors found that ETX based on passive monitoring fails in overloaded (congested) networks. Indeed, a high traffic load (4 packets/s) leads to a congested network so that packets experience many losses. Consequently, a large number of nodes are not able to compute the ETX because they do not receive packets. Hence, routing is interrupted due to a lack of link quality information. This phenomenon leads to a degradation in the network throughput.

The Link inefficiency (LI) proposed by Lal et al. [15] as an approximation of the RNP, is based on passive monitoring and defined as the inverse of the packet success probability (PSP). PSP is an approximated PRR. Lal et al. [15] introduced the PSP metric instead of considering directly the PRR because they assume that accurate PRR measurement requires the reception of several packets, i.e., a large estimation window. This imposes that sensor nodes operate under high duty cycles, which is undesirable for energy-constrained low-power wireless networks. PSP is derived by an analytical expression that maps the average SNR to PSP.

Discussion: It was shown in [19, 28, 39] and even by Lal et al. in [15], that mapping from average SNR to an approximation of PRR may lead to erratic estimation. Hence, using PSP instead of PRR might be unsuitable for link quality estimation.

Four-bit is not only a metric for link quality estimation [23]. It is designed to be used by routing protocols and provides four *bits* of information, compiled from different layers: the *white bit* is from the physical layer and allows to quickly identify good quality links, based on one packet reading. The *ack bit* is from the link layer and indicates whether an acknowledgment is received for a sent packet. The *pin bit* and the *compare bit* are from the network layer and are used for the neighbor table replacement policy. Four-bit assesses link quality as an approximation of the packet retransmissions count. It combines two metrics (i) $estETX_{up}$, as the quality of the unidirectional link from sender to receiver, and (ii) $estETX_{down}$, as the quality of the unidirectional link from receiver to sender. $estETX_{up}$ is exactly the RNP metric, computed based on w_p transmitted/retransmitted data packets. $estETX_{down}$ approximates RNP as the inverse of WMEWMA, minus 1; and it is computed based on w_a received beacons. The combination of $estETX_{up}$ and $estETX_{down}$ is performed through the EWMA filter as follow:

$$four - bit(w_a, w_p, \alpha) = \alpha \times four - bit + (1 - \alpha) \times estETX \qquad (3.5)$$

$estETX$ corresponds to $estETX_{up}$ or $estETX_{down}$: at w_a received beacons, the node derives four-bit estimate by replacing $estETX$ in Eq. (3.5) for $estETX_{down}$. At w_p transmitted/re-transmitted data packets, the node derives four-bit estimate by replacing $estETX$ in Eq. (3.5) for $estETX_{up}$. Four-bit is then both a sender- and received-side LQE and it takes into account link asymmetry. Further, it uses both passive (data packet traffic) and active (beacons traffic) monitoring.

Discussion: To evaluate the performance of Four-bit, Fonseca et al. [23] considered the CTP routing protocol [3]. In CTP, routing consists in building and maintaining a tree towards the sink node, based on link quality estimation. Then, the authors compared the original version of CTP that uses ETX as LQE, against a modified version of CTP that uses Four-bit as LQE. They also involved another routing protocol called MultiHopLQI [47], which also builds and maintains a tree toward the sink node, but LQI is used as LQE. Performance comparison was performed using three metrics: (i) cost, which accounts for the total number of transmissions in the network for each unique delivered packet, (ii) average depth of the routing tree, and (iii) delivery rate, which is the fraction of unique messages received at the root. Fonseca et al. [23] found that CTP based on Four-bit provides better performance (e.g., packet delivery) than the original version of CTP and MultiHopLQI.

The L-NT and L-ETX are two sender-side LQEs that approximate the *RNP* [14]. They are referred as data-driven LQEs because they are based on feedback from unicast data packets. L-NT counts the number of transmissions to successfully deliver a packet then applies the EWMA filter. On the other hand, L-ETX first computes the ratio of the number of acknowledged packets to the total number of transmitted packets based on a certain estimation window. Then, it applies the EWMA filter and inverts the result.

Discussion: Through mathematical analysis and experimental measurements, the authors in [14] demonstrated that L-ETX is more accurate in estimating ETX than L-NT. It is also more stable. However, this result does not mean that L-ETX is accurate at estimating link quality because ETX is not a reference/objective metric. The authors also showed through an experimental study that L-NT, when used as a routing metric, achieves better routing performance than L-ETX, namely a higher data delivery ratio and energy efficiency. This result might be more convincing than the first as it indeed shows that L-ETX is an accurate LQE. Such routing performance can be explained by the fact that L-ETX allows to select stable routes with high quality links.

3.5.3 Score-Based

Some LQEs provide a link estimate that does not refer to a physical phenomena (like packet reception or packet retransmission); rather, they provide a score or a label that is defined within a certain range. In the following, we present an overview on four score-based LQEs: MetricMap [20], WRE [18], F-LQE [41] and CSI [29].

MetricMap is proposed by Wang et al. [20], as an alternative LQE for MintRoute, a hierarchical routing protocol, when the original LQE ETX fails to select routes [2]. Such failure occurs when a node cannot find a route, i.e., a node that can not find a parent (an orphan node) in MintRoute. Wang et al. identified link quality estimation as a classification problem. MetricMap uses a classification algorithm to classify the link among a set of classes (e.g., "Good", "Bad"). This algorithm has as input a feature vector, which consists of a set of metrics that impact link quality, including RSSI, channel load assessment, and node depth. This classification algorithm relies on a training phase, which is performed using a database of training samples. Each sample consists of a feature vector and a corresponding class label.

Discussion: Wang et al. [20] showed that MetricMap combined with ETX improves the network performance in terms of packet delivery rate and fairness. This measures the variability of the delivery rate across all source nodes. However, MetricMap can be used as a back-off metric but not as a sole metric for link quality estimation. This fact is due to the use of learning algorithms, which are greedy algorithms and might be unsuitable to be executed by sensor nodes.

The *Weighted Regression Estimator* (WRE) is proposed by Xu and Lee [18]. They argued that the received signal strength is correlated with distance. This observation was generalized to the fact that a node can determine the quality of the link to its neighbour giving the location of this neighbour. Hence, WRE derives a complex regression function based on an input vector that contains a set of nodes locations together with their links quality known in advance. This function is continuously refined and updated by the knowledge of a new input, i.e., node location and the corresponding link quality. Once derived, this function returns an estimation of the link quality giving the neighbour location.

Discussion: The performance of WRE is evaluated by comparing it to WMEWMA using the same evaluation methodology as that of Woo et al. [2], where PRR is considered as the objective metric. Xu and Lee [18] found that WRE is more accurate than WMEWMA. However, we believe that the introduced estimator is complex and involves computation overhead and high memory storage (due to regression weights determination). Moreover, WRE assumes that link quality is correlated with distance, which is not always true, as proved by several empirical studies on low-power links [9, 10, 48, 49].

The DoUble Cost Field HYbrid (DUCHY) [29] is not a standalone LQE, but a mechanism leading to a routing metric that selects high quality routes with as low a hop count as possible. DUCHY is meant to be integrated within cost-based wireless collection protocols. In fact, it was designed for the control plane of the Arbutus collection protocol [21]. DUCHY sets up a double cost field: an outer field and an inner field. The outer field serves to prescreen potential parents based on their hop count from the sink in order to avoid unnecessarily long routes. It is bootstrapped using hop count information and validated with the RNP count for robustness against asymmetric links.

The inner field serves to select the potential parent leading to the most reliable route. It is initialized with an estimate of the Channel State Information (CSI) based on broadcast control traffic. The CSI estimate is computed using normalized RSSI and LQI, which are both combined into a weighted sum. Both RSSI and LQI are employed because they complement each other: RSSI provides soft information about good links while LQI provides soft information about poor links. As soon as data traffic becomes available, RNP is computed based on link layer acknowledgements and combined with the CSI estimates. Note that, if the link is good, the cost of using the link determined by DUCHY (the inner field) mirrors the CSI estimates. On the other hand, if the link is poor, its cost as computed by DUCHY is dominated by the RNP component. For lossy links, therefore, DUCHY's LQE mechanisms behaves like an RNP-based LQE in spite of its score-based nature. Because it measures the

RNP based on link layer acknowledgements, DUCHY is intrinsically robust against asymmetric links.
Discussion: DUCHY was initially introduced In [29] and integrated within the Arbutus collection protocol for a comparative evaluation where a basic version of Arbutus with DUCHY is shown to outperform MultiHopLQI, another collection protocol that uses LQI as its LQE [47]. In [21], a thorough description of DUCHY as an integral part of Arbutus is provided, and a full-featured version of Arbutus with DUCHY is compared to CTP with LEEP and Four-Bit Link Estimation. These evaluations do not single out the link estimation performance of DUCHY, but encompass its overall performance in the context of routing.

The Triangle Metric [43] combines geometrically the information carried by PRR, LQI, and SNR in order to obtain a fast and reliable estimation. It is built on top of the hardware-based LQEs SNR and LQI, following the observation that the higher their values, the better the quality of a wireless link (see Sect. 3.4). The mean values of SNR and LQI, denoted by $\overline{SNR}$ and $\overline{LQI}$, are represented geometrically on a bi-dimensional Cartesian coordinate system. Thus, the link quality is estimated as the distance of the point $(\overline{SNR}, \overline{LQI})$ from the origin (0,0), i.e., length of hypotenuse c in Fig. 3.3b, which is given by the following expression:

$$d_{\triangle} = \sqrt{\overline{SNR}_w^2 + \overline{LQI}_w^2} \tag{3.6}$$

The Triangle Metric also embeds PRR by computing $\overline{SNR}$ and $\overline{LQI}$ as a "window mean", where the values of SNR and LQI are divided by the total amount of transmitted packets (see Eq. 3.6). This operation is very important, since lost packets carry a significant piece of information about the link quality. Indeed, it may happen that a short-term effect (e.g., multipath fading or interference) would lead to the reception of only one packet out of N with a high SNR and LQI on a link, while a different link would instead receive 5 packets out of N with a mean SNR and mean LQI similar to (or smaller than) the former, with the consequence of having these two links erroneously classified as nearly equivalent. Thus, $\overline{SNR}$ and $\overline{LQI}$ are computed as follows:

$$\overline{SNR} = \frac{\sum_{k=1}^{m} snr_k}{n} \tag{3.7}$$

$$\overline{LQI} = \frac{\sum_{k=1}^{m} lqi_k}{n} \tag{3.8}$$

where n represents the amount of packets sent by a node to sample the channel, m represents the amount of packets that were actually received, and lqi_k and snr_k denote the LQI and SNR of each successfully received packet k.
Discussion: The authors have shown that the Triangle Metric performs well in mobile settings with an observation window of 5–10 packets, getting the best out of each

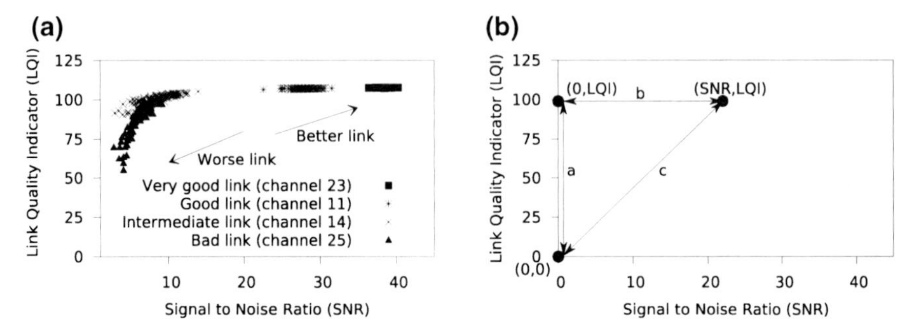

Fig. 3.4 The higher the SNR and the LQI, the better the link quality (**a**). This observation motivates the geometric basis of the Triangle Metric: the link quality can be estimated by computing the distance of the point ($\overline{SNR}$, $\overline{LQI}$) from the origin (0,0), i.e., by calculating the length c of the hypotenuse of the triangle abc (**b**) [43]. **a** Geometrical relationship between SNR and LQI. **b** Computation of the Triangle Metric

metric it is built upon. An interesting task would be to thoroughly compare the performance of other state-of-the-art LQEs in mobile settings, and compare their performance to the ones offered by the Triangle Metric (Fig. 3.4).

The Fuzzy Link Quality Estimator (F-LQE) [41] is a recent estimator, where link quality is expressed as a fuzzy logic rule, which combines desirable link properties, namely the smoothed Packet Reception Ratio (SPRR),[2] link stability factor (SF), link asymmetry (ASL), and channel Signal to Noise Ratio (ASNR). For a particular link, the fuzzy logic interpretation of the rule gives an estimation of its quality as a membership score in the fuzzy subset of good quality links. Scores near 1/0 are synonym of good/poor quality links. Hence, according to FLQE, the membership of a link in the fuzzy subset of good quality links is given by the following equation:

$$\mu(i) = \beta.min(\mu_{SPRR}(i), \mu_{ASL}(i), \mu_{SF}(i), \mu_{ASNR}(i)) \\ + (1-\beta).mean(\mu_{SPRR}(i), \mu_{ASL}(i), \mu_{SF}(i), \mu_{ASNR}(i)) \qquad (3.9)$$

The parameter β is a constant in [0..1]. μ_{SPRR}, μ_{ASL}, μ_{SF}, and μ_{ASNR} represent membership functions in the fuzzy subsets of high packet reception ratio, low asymmetry, low stability, and high channel quality, respectively. All membership functions have piecewise linear forms, determined by two thresholds. In order to get stable link estimates, F-LQE uses EWMA filter to smooth $\mu(i)$ values. F-LQE metric is finally given by:

$$FLQE(\alpha, w) = \alpha.FLQE + (1-\alpha).100.\mu(i) \qquad (3.10)$$

where, $\alpha \ \varepsilon$ [0..1] controls the smoothness and w is the estimation window. F-LQE attributes a score to the link, ranging in [0..100], where 100 is the best link quality and 0 is the worst.

[2] SPRR is exactly the WMEWMA [2].

Discussion: To validate their estimator, the authors of [41] analyzed the statistical properties of F-LQE, independently of higher layer protocols such as MAC collisions and routing. These statistical properties impact its performance, in terms of *reliability* and *stability*. The performance of F-LQE was compared in terms of reliability and stability with 5 existing LQEs: PRR, WMEWMA, ETX, RNP and Four-bit. It was found that F-LQE outperforms all these LQEs because they are only able to assess a single link property. However, F-LQE might involve higher memory footprint and computation complexity as it combines four different metrics capturing four different link properties.

3.6 Conclusion

Link quality estimation in low-power wireless networks is a thorny problem because its accuracy impacts the design and the effectiveness of network protocols. Despite the large number of LQEs reported in the literature, most of them have not been properly evaluated. One of the reasons is the impossibility, or at least the difficulty, to provide a quantitative evaluation of the accuracy of LQEs. In fact, in link quality estimation, there is a lack of a ground-truth metric in relation to which the accuracy of the estimators can be assessed. In classical estimation theory an estimated process is typically compared to a real known process using a certain statistical tool (e.g., least mean square error). However, such comparison is not possible in link quality estimation, since: (1) there is no metric that is considered as the "real" one to represent link quality; and (2) link quality is represented by quantities with different natures, since some estimators are based on the computation of the packet reception ratio (PRR), some others are based on packet retransmission count (RNP), whereas others are hybrid and more complex. Hence an important question arises: *Is it possible to design a benchmark for a thorough performance evaluation of LQEs*?

This question raises two main challenges: The first challenge is to define the performance criteria for the assessment of LQEs. Particularly, given that the accuracy of LQEs cannot be assessed quantitatively due to the lack of an objective metric, it would be conceivable to assess this accuracy qualitatively. The second challenge is to design benchmark scenarios allowing for the computation of LQEs regardless of their nature (e.g., sender-side vs. receiver-side, hardware-based vs. software-based, simple vs. composite). These scenarios define (i) how to establish a rich set of links exhibiting different qualities and (ii) what are the suitable traffic patterns to be exchanged over these links enabling the computation of LQEs based on collected packet-statistics.

The following chapter addresses the above question and provides a thorough performance evaluation study of a set of representative LQEs reported in the literature.

References

1. Jiang P, Huang Q, Wang J, Dai X, Lin R (2006) Research on wireless sensor networks routing protocol for wetland water environment monitoring. In: Proceedings of the 1st international conference on innovative computing, information and control (ICICIC '06). IEEE Computer Society, pp 251–254
2. Woo A, Tong T, Culler D (2003) Taming the underlying challenges of reliable multihop routing in sensor networks. In: Proceedings of the 1st international conference on embedded networked sensor systems (SenSys '03). ACM, pp 14–27
3. Gnawali O, Fonseca R, Jamieson K, Moss D, Levis P (2009) Collection tree protocol. In: Proceedings of the 7th ACM conferefnce on embedded networked sensor systems (SenSys '09). ACM, pp 90–100
4. Li Y, Chen J, Lin R, Wang Z (2005) A reliable routing protocol design for wireless sensor networks. In: Proceedings of the IEEE international conference on mobile adhoc and sensor systems (MASS '05). IEEE Computer Society, pp 61–65
5. Lim G (2002) Link stability and route lifetime in ad-hoc wireless networks. In: Proceedings of the international conference on parallel processing workshops (ICPPW '02). IEEE Computer Society, p 116
6. Koksal C, Balakrishnan H (2006) Quality-aware routing in time-varying wireless networks. IEEE J Sel Areas Commun Spec Issue Multi-Hop Wireless Mesh Networks 24(11):1984–1994
7. Seada K, Zuniga M, Helmy A, Krishnamachari B (2004) Energy-efficient forwarding strategies for geographic routing in lossy wireless sensor networks. In: Proceedings of the 2nd international conference on embedded networked sensor systems (SenSys '04). ACM, pp 108–121
8. Cerpa A, Estrin D (2004) ASCENT: Adaptive self-configuring sEnsor networks topologies. IEEE Trans Mob Comput 3(3):272–285
9. Zhao J, Govindan R (2003) Understanding packet delivery performance in dense wireless sensor networks. In: Proceedings of the 1st international conference on embedded networked sensor systems (SenSys '03). ACM, pp 1–13
10. Cerpa A, Busek N, Estrin D (2003) Scale: A tool for simple connectivity assessment in lossy environments. Tech. rep.
11. Cerpa A, Wong JL, Kuang L, Potkonjak M, Estrin D (2005) Statistical model of lossy links in wireless sensor networks. In: Proceedings of the 4th international symposium on information processing in sensor networks (IPSN '05). IEEE Press, pp 81–88
12. Couto DSJD, Aguayo D, Bicket J, Morris R (2003) A high-throughput path metric for multi-hop wireless routing. In: Proceedings of the 9th annual International conference on mobile computing and networking (MobiCom '03). ACM, pp134–146
13. Kim KH, Shin KG (2006) On accurate measurement of link quality in multi-hop wireless mesh networks. In: Proceedings of the 12th annual international conference on mobile computing and networking (MobiCom '06). ACM, pp 38–49
14. Zhang H, Sang L, Arora A (2010) Comparison of data-driven link estimation methods in low-power wireless networks. IEEE Trans Mob Comput 9:1634–1648. http://doi.ieeecomputersociety.org/10.1109/TMC.2010.126
15. Lal D, Manjeshwar A, Herrmann F (2003) Measurement and characterization of link quality metrics in energy constrained wireless sensor networks. In: Proceedings of the IEEE global telecommunications conference (Globecom '03). IEEE Communications Society, pp 446–452
16. Woo A, Culler D (2003) Evaluation of efficient link reliability estimators for low-power wireless networks. Tech. Rep. UCB/CSD-03-1270, EECS Department, University of California, Berkeley. http://www.eecs.berkeley.edu/Pubs/TechRpts/2003/6239.html
17. Cerpa A, Wong JL, Potkonjak M, Estrin D (2005) Temporal properties of low power wireless links: Modeling and implications on multi-hop routing. In: Proceedings of the 6th international symposium on mobile ad hoc networking and computing (MobiHoc '05). ACM, pp 414–425
18. Xu Y, Lee WC (2006) Exploring spatial correlation for link quality estimation in wireless sensor networks. In: Proceedings of the 4th annual IEEE international conference on pervasive computing and communication (PERCOM '06). IEEE Computer Society, pp 200–211

19. Yunqian M (2005) Improving wireless link delivery ratio classification with packet snr. In: Proceedings of the international conference on electro information technology. IEEE, pp 6–12
20. Wang Y, Martonosi M, Peh LS (2007) Predicting link quality using supervised learning in wireless sensor networks. ACM SIGMOBILE Mob Comput Commun Rev 11(3):71–83
21. Puccinelli D, Haenggi M (2010) Reliable data delivery in large-scale low-power sensor networks. ACM Trans Sen Netw 6(4):1–41
22. Zhang H, Sang L, Arora A (2008) Unravelling the subtleties of link estimation and routing in wireless sensor networks. Tech. rep.
23. Fonseca R, Gnawali O, Jamieson K, Levis P (2007) Four bit wireless link estimation. In: Proceedings of the 6th international workshop on hot topics in networks (HotNets VI). ACM SIGCOMM
24. Baccour N, Koubâa A, Jamaa MB, do Rosário D, Youssef H, Alves M, Becker LB (2011) Radiale: a framework for designing and assessing link quality estimators in wireless sensor networks. Ad Hoc Netw 9(7):1165–1185
25. Kim M, Noble B (2001) Mobile network estimation. In: Proceedings of the 7th annual international conference on mobile computing and networking (MobiCom '01). ACM, pp 298–309
26. Lin S, Zhou G, Whitehouse K, Wu Y, Stankovic JA, He T (2009) Towards stable network performance in wireless sensor networks (rtss '09). In: Proceedings of the 30th IEEE real-time systems symposium. IEEE Computer Society, pp 227–237
27. Baccour N, Koubaa A, Ben Jamaa M, Youssef H, Zuniga M, Alves M (2009) A comparative simulation study of link quality estimators in wireless sensor networks. In: Proceedings of the 17th IEEE/ACM international symposium on modelling, analysis and simulation of computer and telecommunication systems (MASCOTS '09). IEEE, pp 301–310
28. Senel M, Chintalapudi K, Lal D, Keshavarzian A, Coyle EJ (2007) A Kalman Filter Based Link Quality Estimation Scheme for Wireless Sensor Networks. In: Proceedings of the IEEE global telecommunications conference (GLOBECOM '07). IEEE, pp 875–88
29. Puccinelli D, Haenggi M (2008) DUCHY: Double cost field hybrid link estimation for low-power wireless sensor networks. In: Proceedings of the 5th fifth workshop on embedded networked sensors (Hot EmNets'08). ACM
30. Srinivasan K, Dutta P, Tavakoli A, Levis P (2006) Understanding the causes of packet delivery success and failure in dense wireless sensor networks. Tech. Rep. SING-06-00, Stanford Information Networks Group (SING)
31. Srinivasan K, Dutta P, Tavakoli A, Levis P (2010) An empirical study of low-power wireless. ACM Trans Sen Netw 6:1–49
32. Srinivasan K, Levis P (2006) Rssi is under appreciated. In: Proceedings of the 3th workshop on embedded networked sensors (EmNets (2006)
33. Lin S, Zhang J, Zhou G, Gu L, Stankovic JA, He T (2006) Atpc: adaptive transmission power control for wireless sensor networks. In: Proceedings of the 4th international conference on embedded networked sensor systems (SenSys '06). ACM, pp 223–236
34. Mottola L, Picco GP, Ceriotti M, Gună c, Murphy AL (2010) Not all wireless sensor networks are created equal: a çomparative study on tunnels. ACM Trans Sen Netw 7:15:1–15:33
35. Boano CA, Voigt T, Dunkels A, Österlind F, Tsiftes N, Mottola L, Suárez P (2009) Exploiting the LQI variance for rapid channel quality assessment. In: Proceedings of the 8th IEEE international conference on information processing in sensor networks (IPSN), poster session, pp 369–370
36. Tang L, Wang KC, Huang Y, Gu F (2007) Channel characterization and link quality assessment of ieee 802.15.4-compliant radio for factory environments. IEEE Trans Industrial Informatics 3(2):99–110
37. Polastre J, Szewczyk R, Culler D (2005) Telos: Enabling ultra-low power wireless research. In: Proceedings of the 4th international symposium on information processing in sensor networks (IPSN '05). IEEE Press, pp 364–369
38. Zuniga M, Krishnamachari B (2007) An analysis of unreliability and asymmetry in low-power wireless links. ACM Trans Sen Netw 3(2):63–81
39. Aguayo D, Bicket J, Biswas S, Judd G, Morris R (2004) Link-level measurements from an 802.11b mesh network. SIGCOMM Comput Commun Rev 34(4):121–132

40. Gomez C, Boix A, Paradells J (2010) Impact of LQI-based routing metrics on the performance of a one-to-one routing protocol for IEEE 802.15.4 multihop networks. EURASIP J Wirel Commun Netw 2010:6:1–6:20
41. Baccour N, Koubaa A, Youssef H, Ben Jamaa M, do Rosário D, Alves M, Becker, BL (2010) F-LQE: A fuzzy link quality estimator for wireless sensor networks. In: Proceedings of the 7th european conference on wireless sensor networks (EWSN 2010). Springer, pp 240–255
42. Rondinone M, Ansari J, Riihijärvi J, Mähönen P (2008) Designing a reliable and stable link quality metric for Wireless Sensor Networks. In: Proceedings of the workshop on real-world wireless sensor networks (RealWSN). Glasgow, Scotland, pp 6–10
43. Boano CA, Zúñiga MA, Voigt T, Willig A, Römer K (2010) The triangle metric: fast link quality estimation for mobile wireless sensor networks. In: Proceedings of the 19th international conference on computer communications and networks (ICCCN). pp 1–7
44. Qin Y, He Z, Voigt T (2011) Link quality estimation for future cooperating objects. In: Proceedings of the 2nd international workshop on networks of cooperating objects (CONET)
45. Qin Y, He Z, Voigt T (2011) Towards accurate and agile link quality estimation in wireless sensor networks. In: Proceedings of the 10th IFIP annual mediterranean ad-hoc networking workshop (Med-Hoc-Net). pp 179–185
46. Fairhurst G, Wood L (2002) Rfc 3366 : Advice to link designers on link automatic repeat request (arq). Tech. rep.
47. TinyOS MultiHopLQI routing algorithm: http://www.tinyos.net/tinyos-1.x/tos/lib/MultiHopLQI/ (2004)
48. Zuniga M, Krishnamachari B (2004) Analyzing the transitional region in low power wireless links. In: Proceedings of the 1st international conference on sensor and ad hoc communications and networks (SECON '04). IEEE Communications Society, pp. 517–526
49. Reijers N, Halkes G, Langendoen K (2004) Link layer measurements in sensor networks. In: Proceedings of the 1st IEEE international conference on mobile ad-hoc and sensor systems (MASS '04). IEEE Computer Society, pp. 24–27

Chapter 4
Performance Evaluation of Link Quality Estimators

Abstract The objective of this chapter is twofold: First, we present a methodology to evaluate the performance of LQEs regardless their type (e.g., PRR-based versus RNP or score-based, hardware-based vs. software-based). We follow this methodology to evaluate the performance of six representative and well-known LQEs in the literature. The results presented in this chapter should help network protocol designers fully understand the strengths and weaknesses of these LQEs, thus enabling them to make an informed choice.

4.1 Introduction

The accuracy of LQEs greatly impacts the efficiency of network protocols. For instance, many routing protocols rely on link quality estimation to select high quality routes for communication. The more accurate the LQE is, the more correct the decision made by routing protocols in selecting good routes. This is just one example on how important it is to assess the performance of the LQE before integrating it into a particular network protocol. Unfortunately, despite its importance, only a few research efforts addressed the performance evaluation of LQEs in low-power wireless networks [1–4]:

In [1], the authors compared the performance of WMEWMA—their proposed filter-based LQE—with other filter-based LQEs [1]. The comparison has been carried out in terms of (i) reactivity assessed by the settling time and the crossing time, (ii) accuracy evaluated by the mean square error, (iii) stability assessed by the coefficient of variation, and (iv) efficiency assessed by the memory footprint and computation complexity. This comparative study was performed analytically, based on a simple generated trace. The trace generator is based on the assumption that packets transmission corresponds to independent Bernoulli trials. This comparative performance study suffers from the following shortcomings. First and foremost, this study was restricted to filter-based LQEs. Except WMWMA, the involved LQEs are

N. Baccour et al., *Radio Link Quality Estimation in Low-Power Wireless Networks*, SpringerBriefs in Electrical and Computer Engineering,
DOI: 10.1007/978-3-319-00774-8_4,

not well referred in the literature. They were introduced in the study to choose the best filter for link quality estimation. Second, the comparison is based on a simple generated trace, which does not take into account some important characteristics of low-power links. Finally, most of the introduced performance metrics are based on PRR as an objective metric. This is not true since PRR is not the ideal metric for link quality estimation as we will discuss next.

In [2], the main goal was to study the temporal characteristics of low-power links, using a real WSN deployment. The authors compared PRR and RNP in order to select the best metric for link characterization, concluding that RNP is better than PRR. To justify their finding the authors observed different links during several hours, by measuring PRR and RNP every minute. They found that for good-quality and bad-quality links, i.e. according to their definition links having high (>90 %) and low reception rates (<50 %) respectively, PRR follows the same behavior as RNP. However, for intermediate quality links, PRR *overestimates* the link quality because it does not take into account the underlying distribution of packet losses. When the link exhibits short periods during which packets are not received, the PRR can still have high value but the RNP is high so that it indicates the real link state. As a matter of fact, a packet that cannot be delivered may be retransmitted several times before aborting transmission. The authors also analyzed the statistical relationship between RNP and the inverse of PRR by assessing (i) the cumulative distribution function (CDF) of RNP as a function of 1/PRR and (ii) the Consistency level between RNP and 1/PRR. They found that RNP and PRR are not directly proportional.

In [3], the authors compared four-bit to ETX by studying their impact on CTP routing protocol. They considered two versions of CTP, the first uses four-bit as LQE and the second uses ETX as LQE. Then, the authors compared the performance of the two CTP versions. Performance comparison was performed using three metrics: (i) cost, which accounts the total number of transmissions in the network for each unique delivered packet, (ii) average depth of the topology trees, and (iii) delivery rate, which is the fraction of unique messages received at the root. The authors found that CTP based on four-bit provides better performance than that based on ETX.

In the above described works, two evaluation methodologies were proposed:

- Statistical analysis of LQEs [1, 2]: The statistical properties of LQEs (e.g., the standard deviation, the average ...) are analyzed, independently of any external factor that may affect the link quality, such as collisions (each node transmits its data in an exclusive time slot) and routing (a single-hop network).
- Impact of LQEs on the CTP routing protocol [3]: The contribution of the LQE in improving routing performance (e.g., in terms of end-to-end packet delivery) implies on its reliability.

The second evaluation methodology would not be convenient for the performance evaluation of any LQE. The authors of [3] used this approach to compare the performance of ETX and four-bit. The integration of ETX and four-bit in CTP routing is relatively simple as both ETX and four-bit are RNP-based LQEs and CTP routing metric is designed based on RNP-based link quality estimation. However, when a different LQE is considered (e.g., a PRR-based LQE), it becomes necessary to design another routing metric based on this LQE. This situation will complicate the performance analysis of the LQE in question as what is evaluated is not only the LQE but the routing metric based on the LQE.

The first evaluation methodology seems promising as it can be generalized for any type of LQE. However, it is not clear what are the scenarios allowing for the computation of LQEs and also what are the statistics that impact on their performance, especially in what concerns their accuracy/reliability.

The work in [4] addresses these questions and represents a first attempt toward a unified and holistic methodology for the performance evaluation of LQEs. In the rest of this chapter, we consider this evaluation methodology to present a thorough performance evaluation study of a set of representative LQEs in literature. Beforehand, an overview of this evaluation methodology is given.

4.2 Holistic Evaluation Methodology

The evaluation methodology introduced in [4] consists in analyzing the statistical properties of LQEs, independently of any external factor. These statistical properties impact the performance of LQEs, in terms of:

- **Reliability**: It refers to the ability of the LQE to correctly characterize the link state. The reliability of LQEs is assessed *qualitatively*, by analyzing (i) their temporal behavior, and (ii) the distribution of their link quality estimates, illustrated by the scatter plot and the empirical cumulative distribution function (CDF).
- **Stability**: It refers to the ability to resist to transient (short-term) variations (also called fluctuations) in link quality. The stability of LQEs is assessed *quantitatively*, by computing the coefficient of variation (CV) of the link quality estimates.

This evaluation methodology is established in three steps: links establishment, link measurements collection, and data analysis.

4.2.1 Links Establishment

The first step consists in establishing a rich set of links exhibiting different properties, i.e. different qualities. Particularly, it is recommended to have most of links of intermediate quality, i.e., belonging to the transitional region, in order to better evaluate the capability of LQEs. Recall that intermediate quality links are the hardest to assess as they are extremely dynamic and exhibit asymmetric connectivity (refer to Chap. 1 for the description of the three reception regions).

To achieve this goal, the sensor nodes are placed according to a radial layout, as shown in Fig. 4.1, where nodes $N_2 \ldots N_m$ are placed in different circles around a central node N_1. The distance (in meters) between two consecutive circles is denoted as y, and the first circle that is the nearest to N_1 has a radius of x meters.

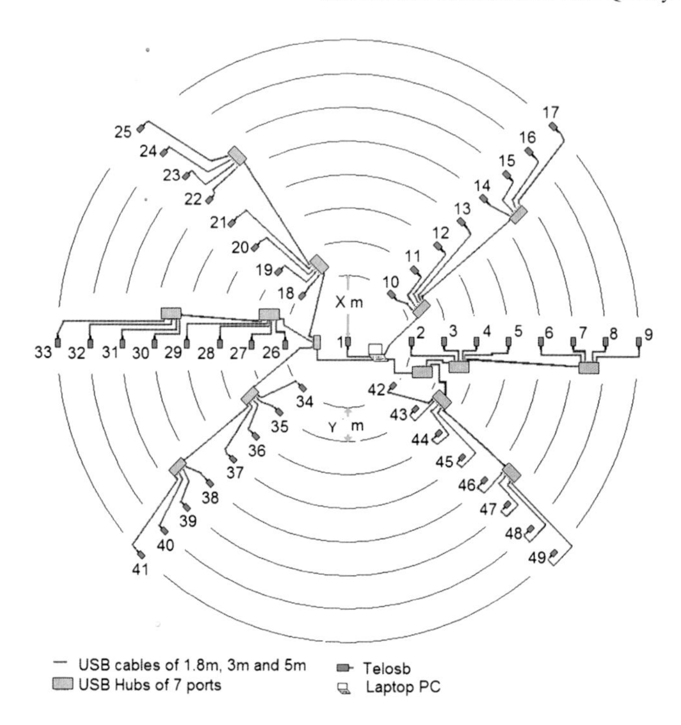

Fig. 4.1 Nodes distribution forming a radial topology

Since distance and direction are fundamental factors that affect the link quality, the underlying links $N_1 \longleftrightarrow N_i$ will have different properties (qualities) by placing nodes $N_2 \ldots N_m$ at different distances and directions from the central node N_1. x and y values should be determined prior to experiments, to have links within the transitional region, which is typically quantified in the literature by means of the PRR.

It is important to note that apart from the network topology (radial topology in our case), network settings also impact the quality of the underlying links. Network settings may include the radio channel, the transmission power, and the environment type (e.g., indoor/outdoor).

4.2.2 Link Measurements Collection

The second step is to create a bidirectional data traffic over each link $N_1 \longleftrightarrow N_i$, enabling link measurements through packet-statistics collection. Packet-statistics collection consists of retrieving statistics, such as packet sequence number, from received and sent packets.

Two traffic patterns are considered: *Burst(N, IPI, P)* and *Synch(W, IPI)*. *Burst(N, IPI, P)* refers to a bursty traffic pattern, where the central node N_1 first sends a burst of packets to a given node N_i. Then, node N_i sends its burst of packets back to N_1. This operation is repeated for P times, where P represents the total number of bursts. A burst is defined by two parameters: N, the number of packets in the burst and IPI, the Inter-Packets Interval in ms. On the other hand, *Synch(W, IPI)* refers to the synchronized traffic, where N_1 and N_i are synchronized to exchange packets in a round-robin fashion. This traffic is characterized by two parameters: IPI and the total number of sent packets, noted by W.

In fact, to accurately assess link asymmetry, it is necessary to collect packet-statistics on both link directions at (almost) the same time. Therefore, the synchronized traffic pattern would be more convenient than the bursty traffic pattern (in particular for large bursts) to evaluate link asymmetry. One other reason to support two traffic patterns is that radio channels exhibit different behaviors with respect to these two traffic patterns, as it will be shown later. In [5], it has been observed that the traffic Inter-Packets Interval (IPI) has a noticeable impact on channel characteristics. For that reason, it is important to understand the performance of LQEs for different traffic configurations.

Exchanged traffic over each link allows for link measurements through packet-statistics collection. Some packet-statistics are evaluated at the receiver side (from received packets) such as global sequence number, time stamp, RSSI, LQI, and background noise. Such data is necessary to compute receiver-side LQEs. On the other hand, sender-side LQEs require other statistics collected at the sender side, such as sequence number, time stamp, packet retransmission count.

4.2.3 Data Analysis

Data analysis comprises two different operations: The first operation is to generate link quality estimates with respect to each LQE, based on packet-statistics collected in the previous step. The second operation corresponds to the statistical analysis of these quality estimates. Concretely, the statistical analysis consist in generating statistical graphics for these LQEs, such as the empirical distribution function and the coefficient of variation, which allows to assess the reliability and the stability of LQEs.

4.3 LQEs Under Evaluation

As reported in the previous chapter, LQEs can be classified as either hardware-based or software-based. Hardware-based LQEs, such LQI, RSSI, and SNR are directly read from the radio transceiver (e.g. the CC2420) upon packet reception. Empirical studies have shown that hardware-based LQEs are not efficient: Despite the fact that they provide a fast and inexpensive way to classify links as either good or bad, they are incapable of providing a fine grain estimation of link quality [3, 6]. Of course, this does not mean that this category of LQEs is useless. Currently, there is a growing awareness that the integration of hardware-based LQEs in software-based LQEs improve the accuracy of the link quality estimation [6–8].

Table 4.1 Characteristics of link quality estimators under evaluation

	Monitoring type	Location	Direction	Class
PRR	Passive	Receiver	Unidirectional	PRR-based
WMEWMA	Passive	Receiver	Unidirectional	PRR-based
RNP	Passive	Sender	Unidirectional	RNP-based
ETX	Active	Receiver	Bidirectional	RNP-based
four-bit	Hybrid	Sender	Bidirectional	RNP-based
F-LQE	Passive	Receiver	Unidirectional	Score-based

This study investigates software-based LQEs. Based on the survey presented in the previous chapter, we have selected most representative LQEs, namely PRR, WMEWMA, ETX, RNP, four-bit, and F-LQE. Table 4.1 presents the most important characteristics of LQEs under evaluation.

4.4 Evaluation Platforms

The performance evaluation of LQEs is conducted with both (i) simulation using TOSSIM 2 simulator [9] and (ii) real experimentation using RadiaLE testbed [4]. An overview of these evaluation platforms is given next.

4.4.1 TOSSIM 2 Simulator

TOSSIM 2.x [9] is an event-driven simulation environment for sensor networks. It is used to simulate the code of real sensor nodes that are implemented using the second release of TinyOS (TinyOS 2.x) [10]. TinyOS 2.x is an operating system and a programming framework developed at UC Berkeley and was specifically designed for sensor networks with small resource capacities. It is written in NesC [11], a C-based language that provides a support for the TinyOS component and concurrency model.

One of the main reasons behind the use of TOSSIM 2 is that it provides an accurate wireless channel model, without which it will not be possible to consider the simulation results as valid [12, 13]. Later on in this thesis, we confirm the accuracy of TOSSIM 2 channel model by conducted the study on LQEs statistical properties, with simulation and real experimentation and comparing the results.

4.4.1.1 Overview of TOSSIM 2 Channel Model

In this section, we present a brief overview on this model. The interested readers can refer to [12, 13] for more details on this channel model.

Basically, the wireless channel model of TOSSIM 2 relies on the *Link layer model* [13] and the *Closest-fit Pattern Matching* (CPM) model [12].

The link layer model of Zuniga et al. [13] corresponds to an analytical model of the PRR according to distance: $PRR(d)$. For non-coherent FSK modulation and Manchester encoding (used by MICAZ motes), this model is given by the following expression:

$$PRR(d) = \left(1 - \frac{1}{2}.exp\left(-\frac{SNR(d)}{2}.\frac{B_N}{R}\right)\right)^{8L} \quad (4.1)$$

where, B_N is the noise bandwidth, R is the data rate in bits, and L is the packet size. These parameters are set to default values. The $SNR(d)$ is given by:

$$SNR(d) = RSS(d) - Pn \quad (4.2)$$

- $RSS(d)$ is the pure (i.e, without noise) received signal strength in dB as a function of distance. It is computed as: $P_t - PathLoss(d)$, where P_t is the transmission power in dB and $PathLoss(d)$ is the path loss in dB as a function of distance. $PathLoss(d)$ corresponds to the *log-normal shadowing* path loss model [13, 14].
- Pn is the sampled noise floor in dB. TOSSIM 2 relies on the CPM model [12] to generate noise floor samples for a given link, which captures the temporal variation of the channel. The principal inputs of this model are the average noise floor at the receiver ($\overline{Pn}$) the noise floor variance, and a noise trace file containing 100 readings.

An important feature of the link layer model is the fact that it takes into account the hardware variance, i.e. the variability of the transmission power among different senders and the variability of the noise floor among different receivers. The hardware variance is the main cause of link asymmetry. To model this variance, the transmission power and the noise floor are considered as Gaussian random variables. Given the variances of the noise floor and the transmission power respectively, the link layer model generates two Gaussian distributions for each variable. Thus, it assigns a transmission power P_t to each simulated sender and a noise floor $\overline{Pn}$, to each simulated receiver. For a given link, P_t is constant over time and $\overline{Pn}$ is used to generate different noise floor readings (i.e. different Pns) to capture the link dynamism.

Now, let's see how TOSSIM 2 uses the channel model presented above: At the beginning of the simulation and based on the channel and radio parameters as well as the topology specification, determined by the user, TOSSIM 2 generates for each link (sender → receiver) the RSS, and the $\overline{Pn}$. TOSSIM 2 models packet reception over a link as a Bernoulli trial with probability equal to PRR. When a packet is received, a simulated receiver samples a noise floor reading (Pn) using the CPM model and computes the PRR according the link layer model (Eq. 4). Then, the receiver node generates a uniform random number (URN). The packet is received (and eventually acknowledged) if the URN is greater than PRR; otherwise it is considered as lost.

4.4.1.2 Advantages and Shortcomings of TOSSIM 2 Channel Model

TOSSIM 2 channel model has the advantage of capturing important low-power links characteristics, namely spatial and temporal characteristics, as well as the asymmetry property. For instance, spatial characteristics are captured by modeling the three reception regions: connected, transitional and disconnected, using the link layer model [13]. To illustrate this fact, the authors in [16] conducted simulations for two environment settings and plotted the PRR as a function of distance, as shown in Fig. 4.2. From this figure, it is possible to observe the three reception regions as observed with real measurements.

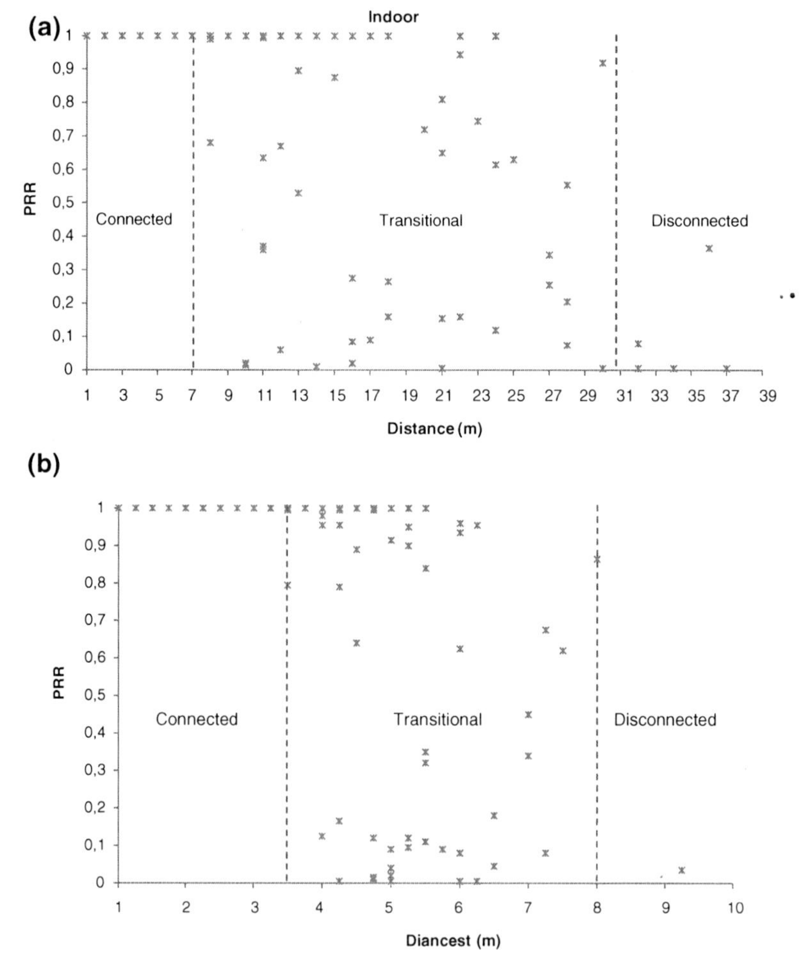

Fig. 4.2 Illustration of TOSSIM 2 channel model reliability: the three reception regions. **a** Indoor environment: aisle of building [15]. **b** Outdoor environment: football field [15]

On the other hand, TOSSIM 2 presents some shortcomings that result from some assumptions. Indeed, TOSSIM 2 uses the log-normal shadowing model to model the path loss. This model has been shown to provide an accurate multi-path channel model. However, it does not take into account the anisotropy property of the radio range, i.e. attenuation of the signal according to the receiver's direction. Therefore, TOSSIM 2 assumes that link quality does not vary according to direction, despite it models the variation according to distance.

Another assumption made by TOSSIM 2 is the fact that $RSS(d)$, which concerns a given link having a distance *d*, is constant over time. This assumption is justified by the fact that the link layer model is designed for static environments [13]. Nevertheless, the "real" received signal strength, which is the $RSS(d)$ added to the noise floor($RSS + Pn$), varies according to time because TOSSIM 2 takes into account the variability of Pn over time using the CPM model [12]. Therefore, link quality (e.g. *RSSI*, PRR, *SNR*) varies over time (for a given link), which captures the link temporal behavior.

4.4.2 RadiaLE Experimental Testbed

Although the performance evaluation of network protocols and mechanisms using network simulators can provide interesting conclusions, there is no guarantee on their validity or trustability level. This fact prompted researchers to build wireless network testbeds for a realistic performance evaluation.

Several testbeds were designed for the experimentation of low-power wireless networks. They can be classified into *general-purpose* testbeds and *special-purpose* testbeds. Most of existing testbeds, including MoteLab [17], Mirage [18], Twist [19], Kansei [20], and Emulab [21] are general-purpose testbeds. They have been designed and operated to be remotely used by several users having different research objectives. On the other hand, special-purpose testbeds, such as Scale [22] and Swat [5] are designed for a specific research objective.

General-purpose testbeds might be not suitable for benchmarking LQEs. Their tendency to cover multiple research objectives prevent them from satisfying some particular requirements. Namely, the physical topology of sensor nodes as well as the environment conditions cannot be managed by the user. However, to assess the performance of LQEs, it is mandatory to design a network topology, where the underlying links are of different qualities. Especially, it is highly recommended to have links with moderate quality and dynamic behaviour.

Many researchers develop their own tesbeds to achieve *a specific goal*. These belong the category of special-purpose testbeds. To our best knowledge, none of the existing testbeds was devoted for the performance evaluation of LQEs. Some testbeds have been dedicated for link measurements, such as SCALE [22] and SWAT [5], but they were exploited for analyzing low-power link characteristics rather than the performance evaluation of LQEs. Some well-known testbeds for low-power wireless networks, belonging to this category are described in what follow:

SCALE [22] is a tool for measuring the Packet Reception Ratio (PRR) LQE. It is built using the EmStar programming model. Each sensor node runs a software stack, allowing for sending and receiving probe packets in a round robin fashion, retrieving packet-statistics, and sending them through serial communication. All Sensor nodes are connected to a central PC via serial cables and serial multiplexors. The PC runs different processes—one for each node in the testbed—that perform data collection. Based on the collected data, other processes running on the PC allow for connectivity assessment through the derivation of the PRR of each unidirectional link. Thus, the network connectivity can be visualized during the experiment runtime.

SWAT [5] is a tool for link measurements. The supported link quality metrics (or LQEs) include PRR and hardware-based metrics: *RSSI*, *LQI*, noise floor, and *SNR*. SWAT uses the same infrastructure as SCALE: sensor nodes (MICAZ or TelosB) are connected through serial connections or Ethernet to a central PC. SWAT provides two user-interfaces (UIs), written in HTML and PHP. Through the HTML UI, users can specify the experiment parameters. The interface invokes Phyton scripts to ensure host-mote communication for performing specific operations, namely sending commands to motes (to control them) and storing raw packet-statistics retrieved from motes into a database. The PHP UI is used to set-up link quality metrics, and to collect some statistics such as PRR over time and correlation between PRR and *RSSI*. Then the UI invokes Phyton scripts to process the collected data and display reports.

SCALE is compatible with old platforms (MICA 1 and MICA 2 motes) which do not support the *LQI* metric. This metric has been shown as important to understand and analyze channel behavior in low-power wireless networks [23]. On the other hand, SWAT is not practical for large-scale experiments, as some configuration tasks are performed manually. Both SWAT and SCALE allow for link measurements through packet-statistics collection but the collected data do not enable to compute various LQEs, namely sender-side LQEs, such as four-bit [3] and RNP [2]. The reason is that SWAT and SCALE do not collect sender-side packet-statistics (e.g. number of packet retransmissions).

The testbeds described above use one-Burst traffic, where each node sends a burst of packets to each of their neighbours then passes the token to the next node to send its burst. This traffic pattern cannot accurately capture the link *Asymmetry* property as the two directions (uplink and downlink) will be assessed in separate time windows. Thus, traffic patterns that improve the accuracy of link Asymmetry assessment are mandatory. In addition, as it was observed in [5], the traffic Inter-packets Interval has a noticeable impact on channel characteristics. For that reason, it is important to understand the performance of LQEs for different traffic configurations/patterns.

RadiaLE [4] is a recent testbed that allows to evaluate the performance of LQEs based on the evaluation methodology presented in Sect. 4.2, i.e., by analyzing the statistical properties of LQEs. Especially, RadiaLE solves the above mentioned deficiencies of existing experimental testbeds and presents the following advantages:

- Provides abstractions to the implementation details by enabling its users to configure and control the network, as well as analyzing the collected packet-statistics database, using user-friendly graphical interfaces.
- Due to the flexibility and completeness of the collected database, a wide range of LQEs can be integrated in RadiaLE.
- Supports two traffic patterns, *Bursty* and *Synchronized*, having different parameters that can be tuned by the user in the network configuration step.
- The RadiaLE software is publicly available as an open-source at [24], together with all relevant information and supporting documentation (e.g. installation and user guides).

Next, we provide an overview of RadiaLE. Especially, we give some implementation details in what concern RadiaLE hardware and software components as well as traffic patterns.

4.4.2.1 Hardware Components

The hardware architecture of RadiaLE, roughly illustrated in Fig. 4.3a, involves three main components: the sensor nodes, the USB tree, and the control station (e.g. laptop PC).

- **Sensor nodes**: The sensor nodes are programmed in nesC [11] over TinyOS 2.x [25]. They do not rely on a particular communicating technology such as Zigbee or 6LowPAN. They also do not use any particular protocol at MAC and network layers. In fact, traffic patterns are designed so that collisions are avoided; also a single-hop network is deployed. The goal of these design choices is analyze the statistical properties of LQEs independently any external factor.
  In the experiments with RadiaLE, TelosB motes [26] are deployed, which are equipped with IEEE 802.15.4 radio compliant chip, namely the CC2420 radio chip [27]. Other platforms (e.g., MICAz) and other radio chip (e.g., CC1000) can also be used with RadiaLE framework. This requires some minor modifications at RadiaLE software tool (specifically, the Experiment Control Application and the nesC application). In fact, if users use platforms other than TelosB but based on the CC2420 radio chip, modifications should only concerns the computation of the sensing measures (e.g., temperature, humidity, and light). On the other hand, if users use different platforms based on other radio chip than the CC2420, additional modifications concerning RSSI and LQI reading, and channel setting should be carried out.
- **USB tree**: The motes are connected to a control station (PC) via a combination of USB cables and *active* USB hubs constituting a USB tree. This USB tree is used as a logging/control reliable channel between the motes and the PC.
  Using *passive* USB cables, serial data can only be forwarded over distances that do not exceed 5 meters. RadiaLE uses *active* USB hubs, daisy-chained together, depending on the distance between the sensor node and the PC (refer to Fig. 4.3), in order to forward serial data over large distances. Active USB hubs are also useful

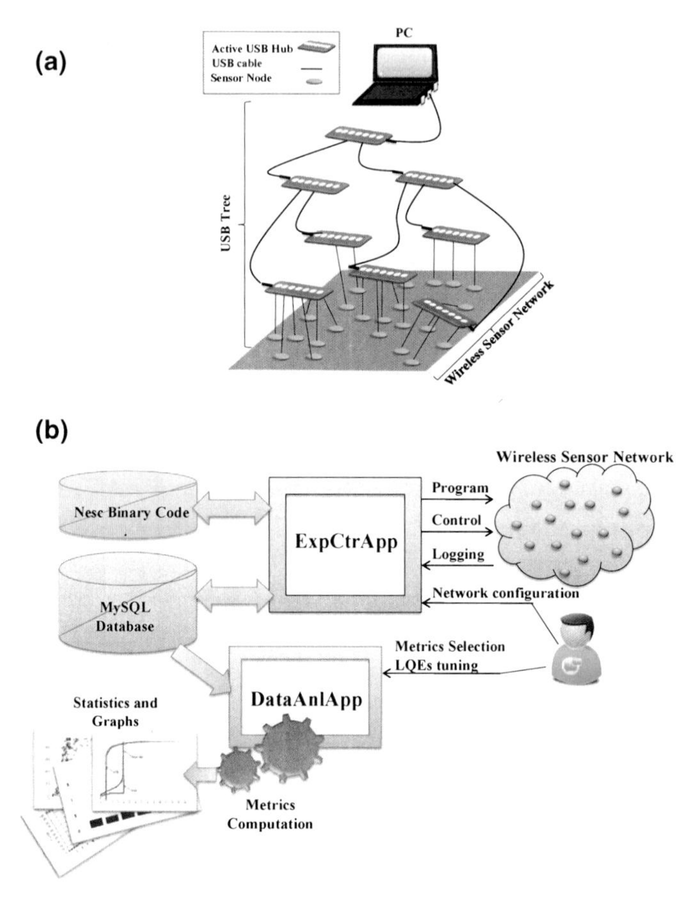

Fig. 4.3 Testbed hardware and software architectures. **a** Hardware architecture. **b** Software architecture

to connect a set of devices (motes or other USB hubs) as shown in Fig. 4.3, and provides motes with power supply.

4.4.2.2 Software Components

RadiaLE software tool contains two independent applications: Experiment Control java application (ExpCtrApp), and Data Analysis Matlab application (DataAnlApp) (Figs. 4.4 and 4.5).

- **The experiment Control application (ExpCtrApp)** It provides user interfaces to ensure multiple functionalities, namely motes programming/control, network configuration and data logging into a MySQL database (Fig. 4.5). These functionalities are described next.

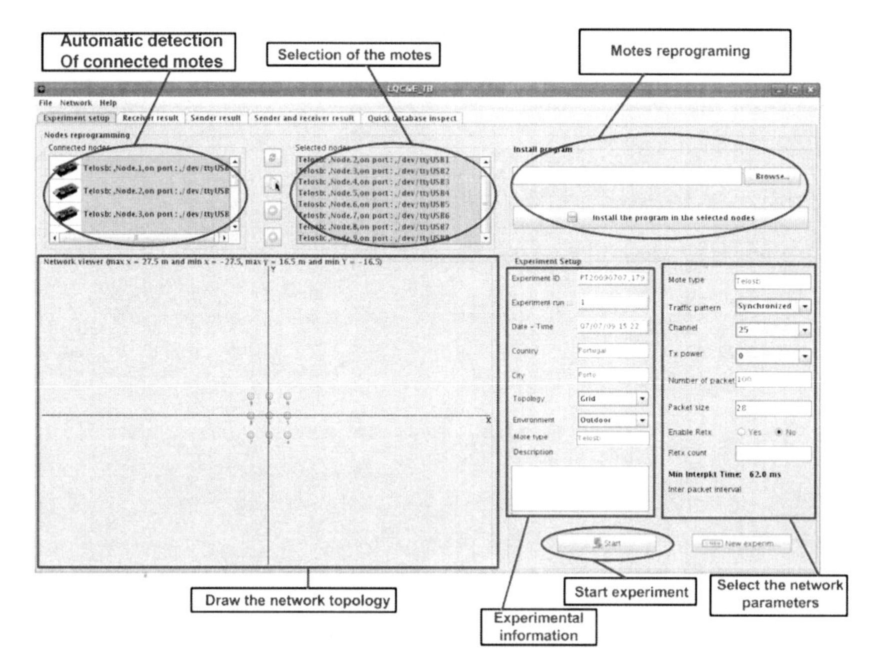

Fig. 4.4 ExpCtrApp Java application main functionalities

Motes programming: A nesC application defines a set of protocols for any bidirectional communication between the motes and between the motes and the ExpCtrApp. The ExpCtrApp automatically detects the motes connected to the PC and programs them by installing the nesC application binary code.
Network configuration: The ExpCtrApp enables the user to specify network parameters (e.g. traffic pattern, packets number/size, inter-packet interval, radio channel, transmission power, link layer retransmissions on/off and max. count). These settings are transmitted to the motes to start performing their tasks.
Link measurements gathering: Motes exchange data traffic in order to collect packet statistics such as sequence number, *RSSI*, *LQI*, *SNR*, time stamp or background noise, which are sent via USB to the ExpCtrApp in the PC, which stores these log data into a MySQL database.
Motes control: ExpCtrApp exchanges commands with the motes to control data transmission according to the traffic pattern set at the network configuration phase. Figure 4.6 illustrates the implementation of the bursty and synchronized traffics. Particularly, this figure shows the interaction between the PC (i.e. ExpCtrApp) and two motes constituting the link $N_1 \longleftrightarrow N_i$, through commands exchange. The ExpCtrApp also provides: (i) a *network viewer* to visualize the network map and the link quality metrics (e.g. PRR, *RSSI*) in real-time; and (ii) a *database inspector* to view raw data retrieved from the motes in real-time.

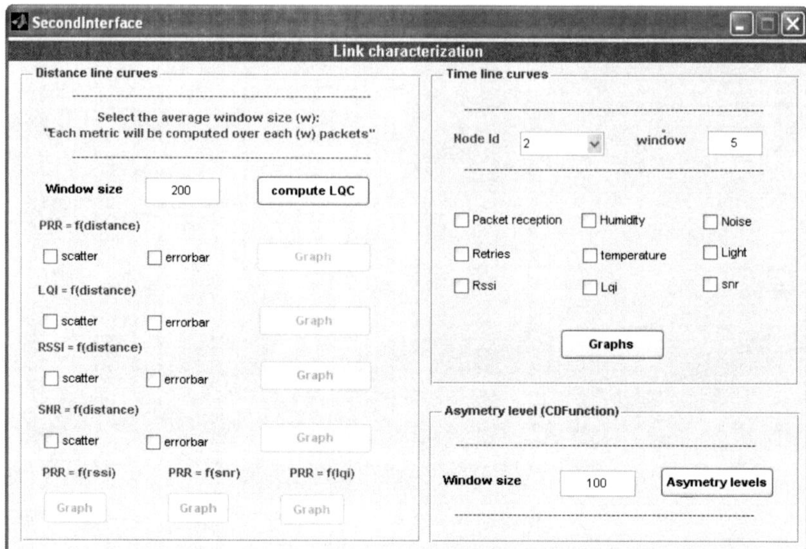

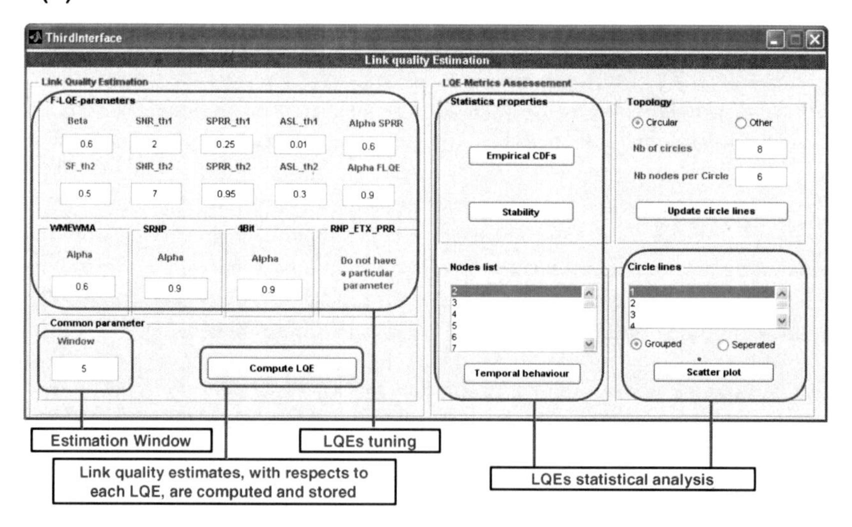

Fig. 4.5 DataAnlApp Matlab application main functionalities. **a** Links characterization interface. **b** Link quality estimation interface

- **The data analysis application (DataAnlApp)** The DataAnlApp application provides user interfaces that connects to the database, and process data to ensure two major functionalities (Fig. 4.5).
  The first functionality provides a set of configurable and customizable graphs that help understanding the channel behaviour, as illustrated in Fig. 4.5a.

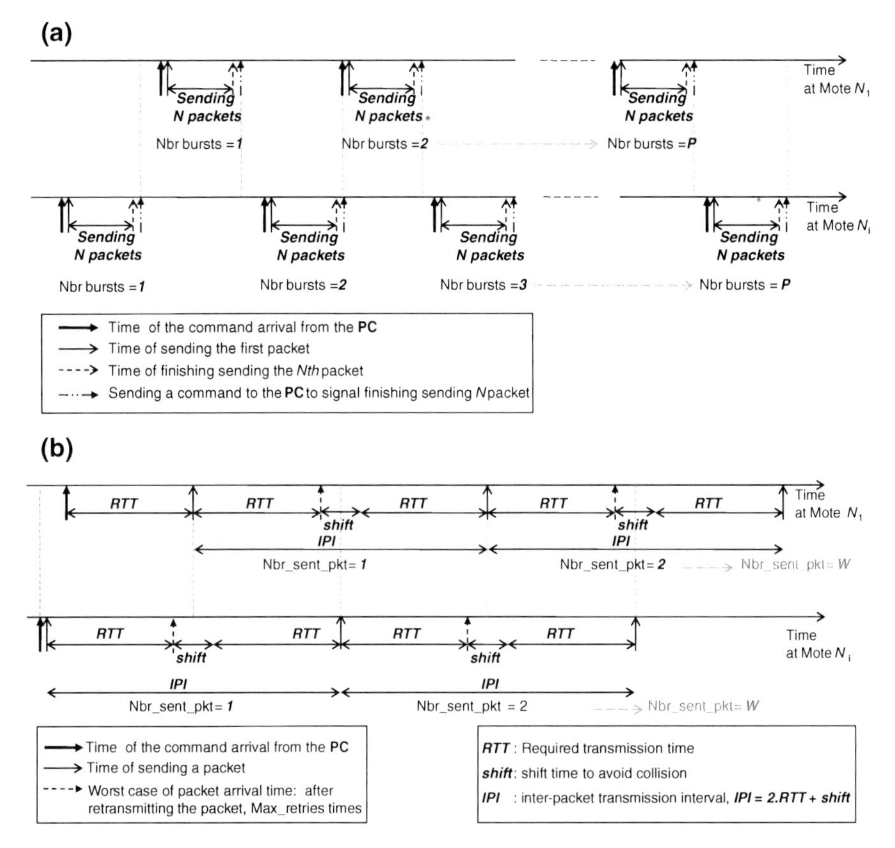

Fig. 4.6 Interaction between mote N_1, mote N_i and the PC, allowing for a Bursty or Synchronized traffic exchange between the two Motes. When N_1 and N_i finish their transmission, the PC triggers a new Bursty or Synchronized traffic exchange between N_1 and N_{i+1}. **a** *Burst(N, IPI, P)* traffic. After receiving the command from the PC, the mote sends a burst of N packets to the other mote, with an inter-packet interval equal to *IPI* ms. This operation is repeated until reaching a total number of sent bursts equal to P. **b** *Synch(W, IPI)* traffic. After receiving the command from the PC, the mote sends a packet to the other mote each *IPI* ms, until reaching a total number of sent packets equal to W

The second functionality provides an assistance to RadiaLE users to evaluate the performance of their estimators, as illustrated in Fig. 4.5b. DataAnlApp proposes a set of LQEs that are configured and computed off-line (i.e., after the experiment finishes), based on the collected data available in the MySQL database. In fact, this constitute one of the interesting features of RadiaLE as it enables to perform the statistical analysis of a given LQE under different settings/configurations without the need to repeat experiments. Further, new LQEs can be integrated to the DataAnlApp and also validated without the need to repeat the experiment. Note the integration of any new LQE is possible thanks to the flexibility and completeness of the collected empirical data.

Once LQEs are computed, the DataAnlApp provides pertinent graphs to visualize their statistical properties and deduce their performance in terms of reliability and stability. Currently, DataAnlApp integrates most well-known LQEs (e.g., PRR, WMEWMA, ETX, RNP and four-bit).

4.4.2.3 Traffic Patterns

An important feature that distinguishes RdiaLE from existing testbeds is the fact that it provides two traffic patterns: Bursty and Synchronized (refer to Sect. 4.2.2 for their description). Figure 4.6 shows some implementation details illustrating the interaction between mote N_1, mote N_i and the PC, allowing for a Bursty or Synchronized traffic exchange between the two Motes.

4.5 Performance Analysis

In this section, we present a comparative performance evaluation study of the considered LQEs, i.e, PRR, WMEWMA, ETX, RNP, four-bit, and F-LQE based on the evaluation methodology described in Sect. 4.2 and using TOSSIM 2 and RadiaLE as evaluation platforms. LQEs are set as follow: the a history control factor $\alpha = 0.9$ for four-bit, as suggested in [3] and $\alpha = 0.6$ for WMEWMA, as suggested in [1] and an estimation window $w = 5$.

For the fairness and brevity of this section, we start by presenting the experimental study and discussing the related results in detail. Then, we give a brief overview on the simulation study, as most simulation results are shared with experimental results.

4.5.1 Experimental Study

4.5.1.1 Experiments Description

In our experiments, we have deployed a single-hop network with 49 TelosB motes distributed according to the radial topology (refer to Fig. 4.7 and Sect. 4.2.1), where x varies in $\{2, 3\}$ meters and y is equal to 0.75 meter. Figure 4.7 shows the topology layout of the 49 motes at an outdoor environment (garden in the ISEP/Porto). Note that x and y were pre-determined through several experiments, prior to deployment. In each experiment, we set x and y to arbitrary values. At the end of the experiment, we measured the average PRR for each link. The chosen x and y are retained if the average PRR, with respect to each link, is between 90 and 10 %. This means that the underlying links intermediate quality and therefore belong to the *transitional* region. As we have mentioned before, the transitional region is the most relevant context to assess the performance of LQEs. It can be identified by analyzing the average PRR

Fig. 4.7 Deployment of the Radial topology at an outdoor environment

of the link [2, 28]). Note that the average PRR of a given link is the average over different PRR samples. Each PRR sample is computed based on w received packets, where w is the estimation window.

Using RadiaLE ExpCtrApp software, we performed extensive experimentations through different sets of experiments. In each experiments set, we varied a certain parameter to study its impact, and for each parameter modification the experiment was repeated. Parameters under consideration were traffic type (3 sorts of bursty traffic and 1 synchronized traffic), packet size (28/114 bytes), channel radio (20/26), and the maximum retransmissions count (0/6). The duration of each experiment was approximately 8 hours. Table 5.1 depicts the different settings for each experiments set. The transmission power was set to the minimum, -25 dBm, in order to reach the transitional region (i.e. have all links with moderate connectivity) at shorter distances. At the end of the experiments, we used DataAnlApp, the RadiaLE data analysis tool, to process packets-statistics retrieved from each bidirectional link $N_1 \longleftrightarrow N_i$ and stored in a database, which results in LQEs computation and the statistical graphs generation.

4.5.1.2 Experimental Results

In this section, we present the experimental results related to the performance comparison of PRR, WMEWMA, ETX, RNP, four-bit, and F-LQE in terms of reliability and stability (refer to Sect. 4.2 for the definition of these criteria).

We point out that we collected empirical data from the 48 links of our Radial topology. Furthermore, we repeated the experiments twice; for $x = 2$ and $x = 3$. In total, we obtained empirical data from $48 * 2 = 96$ bidirectional links. We have considered all these links to conduct our statistical analysis study, namely the empirical CDF and the CV with respect to each LQE (e.g., in Figs. 4.8 and 4.11). Considering all these links together is important for the following reasons: (i) it improves the accuracy

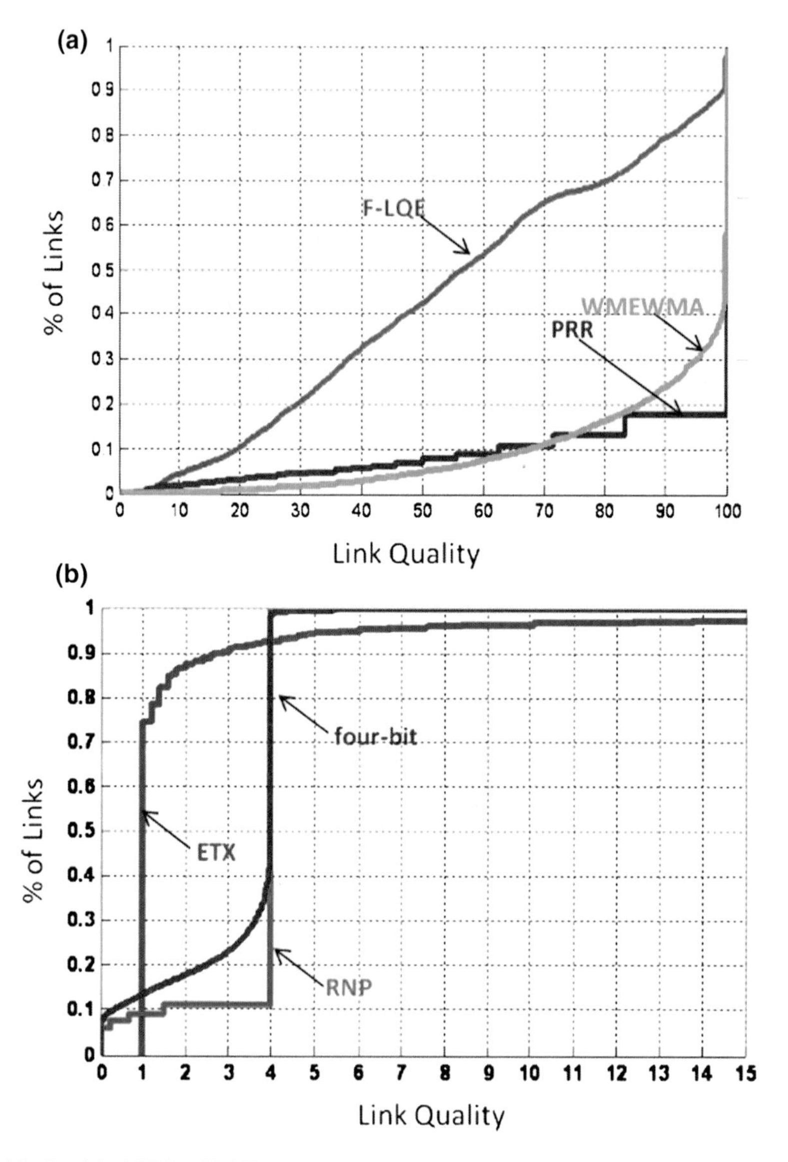

Fig. 4.8 Empirical CDFs of LQEs, based on all the links in the network (default setting)

of our statistical analysis by considering a large sample set and (ii) it avoids having the statistical analysis being biased by several factors such as distance and direction, which provides a global understanding of LQEs behavior. In contrast, regarding the evolution of LQEs in space (e.g., in Fig. 4.9) or in time (e.g., in Fig. 4.10), the observation is made for a particular representative link, because considering all links is not relevant as it was the case with the CDF and CV.

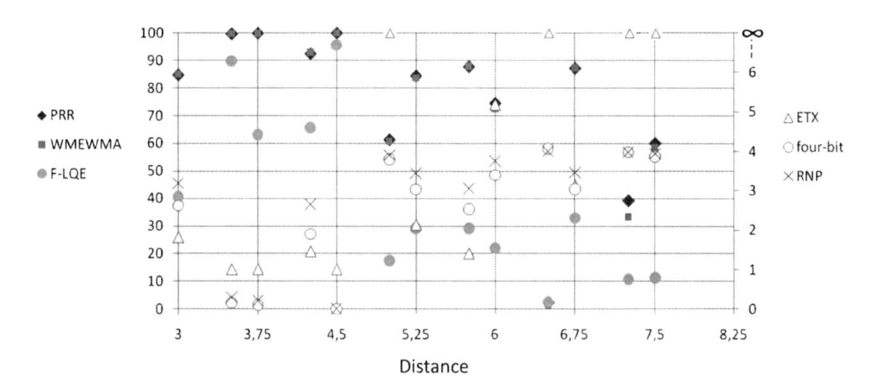

Fig. 4.9 Scatter plot of each LQE according to distance in meters (default setting). Note we subtract 1 from ETX, to account only for the retransmitted packets

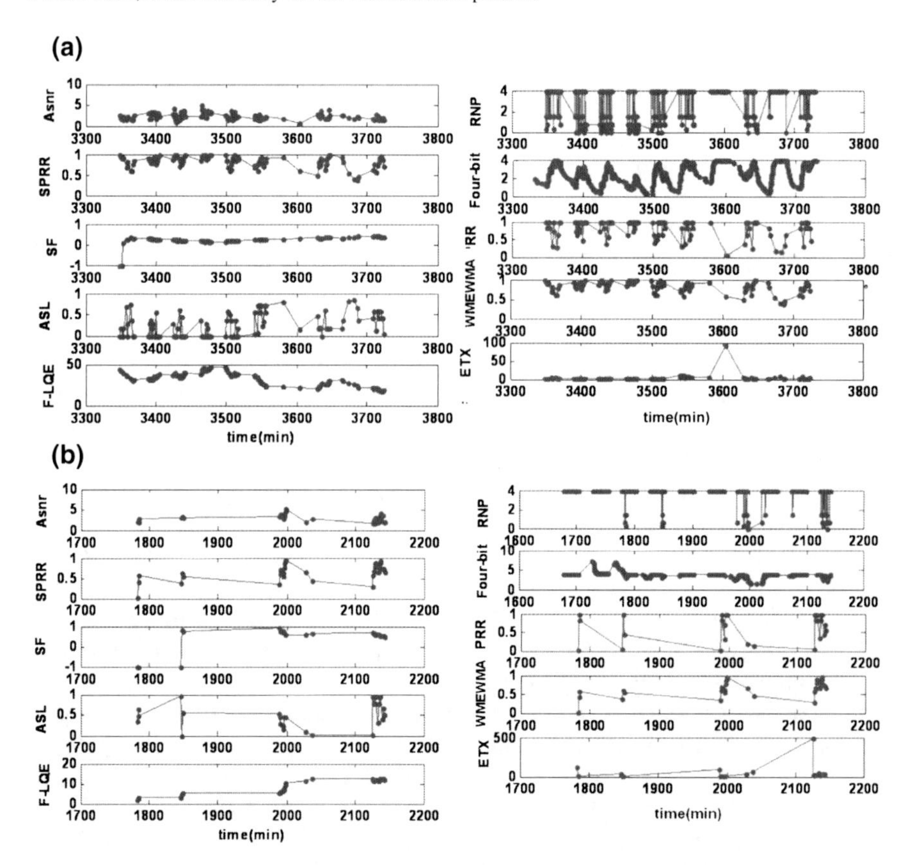

Fig. 4.10 Temporal behaviour of LQEs when faced with links with different qualities (refer to Table 4.2—Scenario 5). **a** Moderate link. **b** Bad link

Reliability

Figure 4.8 presents the global Fig. 4.8 presents the global empirical CDFs of all LQEs. This figure shows that PRR, WMEWMA, and ETX, which are PRR-based LQEs, overestimate the link quality. For instance, this figure shows that almost 80 % of links in the network have a PRR and WMEWMA greater than 84 % (which is considered a high quality value). Also 75 % of the links have ETX equal to 1, (i.e. 0 retransmissions, which also means high quality). The reason of this overestimation is the fact that PRR-based LQEs are only able to evaluate the link delivery, and they are not aware of the number of retransmissions made to deliver a packet. A packet that is lost after one retransmission or after n retransmissions will produce the same estimate. On the other hand, Fig. 4.8 shows that four-bit and RNP, which are RNP-based, underestimate the link quality. In fact Fig. 4.8 shows that almost 90 % of the links have RNP equal to 4 retransmissions (maximum value for RNP), which means that the link is of very bad quality. We observe that Four-bit provides a more balanced characterization of the link quality than RNP, since its computation also accounts for PRR. This underestimation of RNP and four-bit is due to the fact that they are not able to determine if these packets are received after these retransmissions or not. This discrepancy between PRR-based and RNP-based link quality estimates is justified by the fact that most of the packets transmitted over the link are correctly received (high PRR) but after a certain number of retransmissions (high RNP). More importantly, each of these LQEs assess a single and different link property (either packet reception or number of packet retransmission). As for F-LQE, Fig. 4.8 shows that the distribution of link quality estimates is nearly an uniform distribution, which means that F-LQE is able to distinguish between links having different link qualities. In other words, F-LQE neither overestimates the link quality like PRR-based estimators do, nor underestimates it like RNP-based estimators do. It takes into account different properties of radio links, namely Reception Ratio, stability, asymmetry, and channel quality, in order to provide a global characterization of the real link state.

These observations are confirmed by Figs. 4.9 and 4.10. Figure 4.9 illustrates the difference in decisions made by LQEs in assessing link quality. For instance, at a distance of 6 m, PRR and WMEWMA assess the link to have moderate quality (74 and 72 % respectively), whereas RNP and four-bit assess the link to have poor quality (around 3.76 retransmissions). At a distance of 6 m, ETX is PRR-based, but in contrary to other PRR-based LQEs, it assesses the link to have poor quality (5 retransmissions). The reason is that the PRR in the other direction is low (refer to Eq. 2). Figure 4.9 shows also that F-LQE estimates are more scattered than those of the other link estimators, which means that F-LQE is able to provide a fine grain classification of links comparing to the other LQEs.

PRR, WMEWMA, ETX and F-LQE are computed at the receiver side, whereas RNP and four-bit are computed at the sender side. When the link is of a bad quality, the case of the link in Fig. 4.10b, packets are retransmitted many times without being able to be delivered at the receiver. Consequently, receiver side LQEs can not be updated and they are not responsive to link quality degradation. On the other hand, sender side LQEs are more responsive. This observation can be clearly understood from Fig. 4.10b.

In summary, traditional LQEs, including PRR, WMEWMA, ETX, RNP and four-bit have been shown not sufficiently reliable, as they either overestimate or underestimate link quality. On the other hand, F-LQE, a more recent LQE has been shown more reliable as it provides a fine grain classification of links. However, F-LQE as well as PRR, WMEWMA and ETX are not responsive to link quality degradation because they are receiver-side LQEs. RNP and four-bit are more responsive as they are computed at the sender side.

Stability

A link may show transient link quality fluctuations (Fig. 4.10) due to many factors mainly related to the environment, and also to the nature of low-power radios, which have been shown to be very prone to noise. LQEs should be robust against these fluctuations and provide stable link quality estimates. This property is of a paramount importance in low-power wireless networks. For instance, routing protocols do not have to recompute information when a link quality shows transient degradation, because rerouting is a very energy and time consuming operation.

To reason about this issue, we measured the sensitivity of the LQEs to transient fluctuations through the coefficient of variation of its estimates. Figure 4.11 compares the sensitivity (stability) of LQEs, with respect to different settings (refer to Table 4.2). According to this figure, we retain the following observations. First, generally, *F-LQE* is the most stable LQE. Second, WMEWMA is more stable than PRR and four-bit is more stable than RNP. The reason is that WMEWMA and four-bit use filtering to smooth PRR and RNP respectively. Third, except ETX, PRR-based LQEs, i.e. PRR and WMEWMA, are generally more stable than RNP-based LQEs, i.e. RNP and four-bit. ETX is PRR-based, yet it is shown as unstable. The reason is that when the PRR tends to 0 (very bad link) the ETX will tend to infinity, which increases the standard deviation of ETX link estimates.

4.5.2 Simulation Study

In this section, we examine the statistical properties of LQEs using TOSSIM 2 simulation. Importantly, we compare the simulation results to the experimental results presented in the previous section in order to assess the reliability of TOSSIM 2.

4.5.2.1 Simulations Description

As reported in Sect. 4.4.1.1, TOSSIM 2 assumes that link quality varies according to distance, but it does not vary according to direction. To cope with this limitation, we have considered the radial topology while eliminating the *direction* option, which results in a linear topology of nodes $N_1 \ldots N_m$, where the y parameter refers to

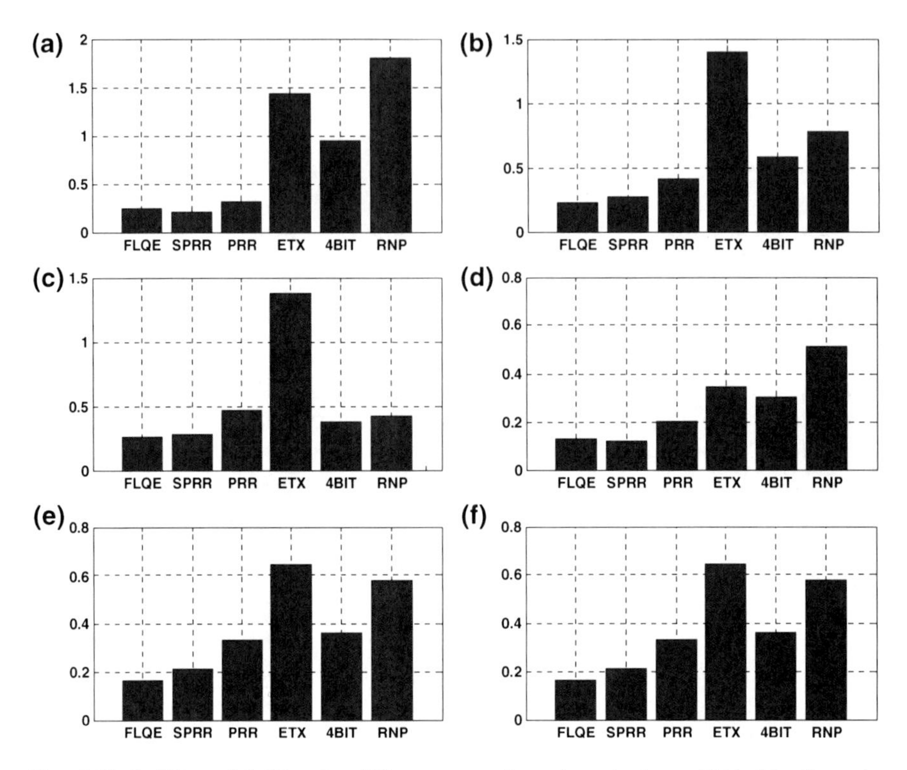

Fig. 4.11 Stability of LQEs, for different network settings (refer to Table 4.2—Scenarios $1, 2, \ldots 5$). **a** Default settings. **b** Packet size : 114 Bytes. **c** Channel 20. **d** Burst(100,1000,2). **e** Burst(200,500,4). **f** Synch(200,1000)

the distance between two consecutive nodes except that between N_1 and N_2, which corresponds to the x parameter.

Giving that, we have considered a single-hop network of 10 sensor nodes (i.e., $m = 10$), where y is fixed to 1 m, and x varies in the set $\{1, 8, 9.25, 10.5, 11.75, 17, 18.25, 19.5, 19.75, 27\}$. We have chosen x values based on a prior study, where we analyzed the three reception regions in both indoor and outdoor environments. The result of this study can be illustrated in Fig. 4.2. Hence, these particular values of x leads to a rich set of links, and copes with the limitation of TOSSIM 2 channel model.

The nodes do not rely on a particular communicating technology such as Zigbee or 6LowPAN. They also do not use any particular protocol at MAC and network layers.

The simulated network is set to an Indoor environment configuration, as described in [15] (The same simulation results have been found for an Outdoor environment configuration). The maximum retransmissions count is set to 6. Other parameters, such as the transmission power are kept to default values as TOSSIM 2 do not permit to tune them. Regarding the traffic pattern, nodes exchange a bursty traffic: Burst(400,720,6). Based on exchanged traffic, nodes perform link measurements

Table 4.2 Experiment scenarios

	Traffic Type	Pkt Size	Channel	Rtx count
Scenario 1: Impact of traffic	{Burst(100,100,10), Burst(200,500,4), Burst(100,1000,2), Synch(200,1000)}	28	26	6
Scenario 2: Impact of Pkt size	Burst(100,100,10)	{28, 114}	26	6
Scenario 3: Impact of channel	Burst(100,100,10)	28	{20, 26}	6
Scenario 4: Impact of Rtx count	Burst(100,100,10)	28	26	{0, 6}
Scenario 5: Default settings	Burst(100,100,10)	28	26	6

Burst(N, IPI, P) and Synch(W, IPI); N: Number of packets per burst, IPI: inter-packets interval, P: number of bursts, W: total number of packets

through packet-statistic collection. LQEs are implemented at the nodes application level (Source codes are available in [24]) and computed based on the collected link measurements.

At the end of the simulation, we gather a trace file that contains, for each LQE, the different link quality estimates computed at each w. By processing the simulation trace file using a software tool similar to the DataAnlApp, the different statistical graphs are generated.

4.5.2.2 Simulation Results

In this section, we present the simulation results related to the performance comparison of PRR, WMEWMA, ETX, RNP, and four-bit, in terms of reliability and stability.

Reliability

It can be clearly observed that the empirical CDF of LQEs, computed based on all links in the simulated networks and illustrated in Fig. 4.12, has the same shape as the empirical CDF of LQEs computed based on real experiments (Fig. 4.8). Consequently, it can be confirmed, based on these simulation results, that PRR, WMEWMA, and ETX over-estimate the link quality. RNP and four-bit under-estimate the link quality. On the other hand, F-LQE has a uniform distribution. Moreover, RNP and four-bit are computed at the sender side and are more responsive to link quality degradations. This fact can also be observed from the temporal behavior depicted in Fig. 4.13.

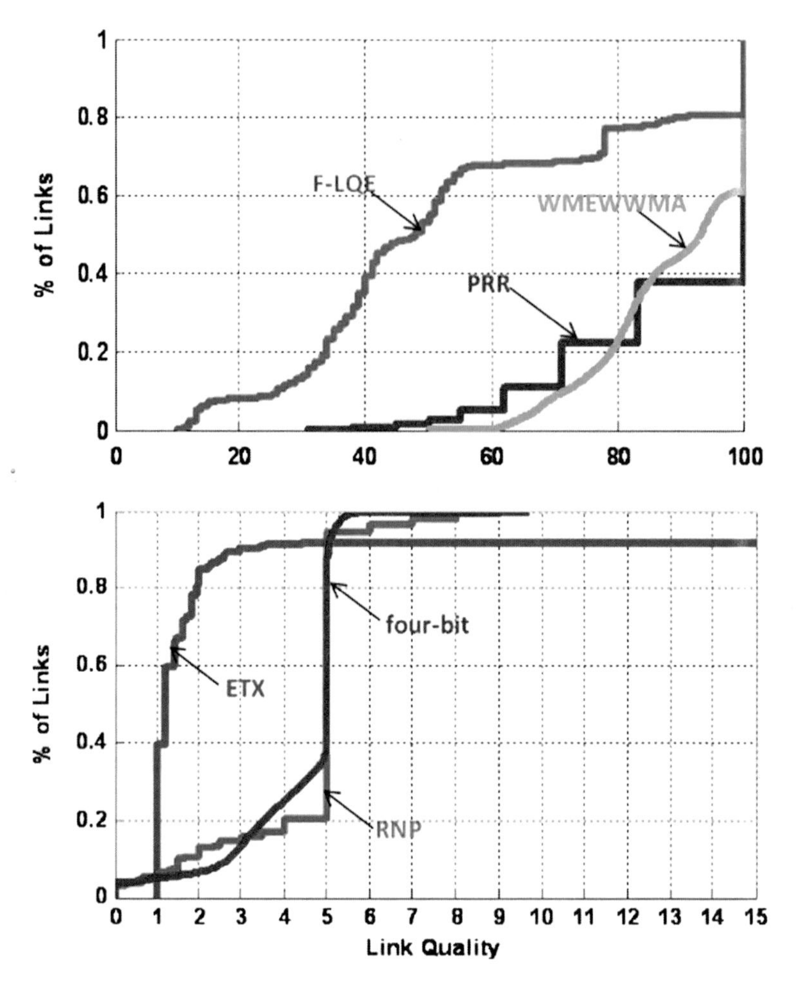

Fig. 4.12 Empirical CDFs of LQEs, based on all the links in the simulated network, and observed by Tossim 2 simulation

Stability

Figure 4.14 shows that RNP and four-bit are more instable than PRR, WMAWMA and F-LQE, as they are more responsive to link quality fluctuations. This finding confirms the results found in the experimental study (Fig. 4.11). However, ETX is shown to be much more instable in the experimental study than in simulation. The instability of ETX in the experimental study is due to the presence of very low PRRs (in the range of 10^{-3}). On the other hand, in simulation, PRR rarely takes low values. This should be due to the assumption that packet reception is a Bernoulli trial, and also to the non-ideality of random number generators. Nevertheless, it is well-known that simulation can not provide very accurate models, as very accurate models will be at the cost of high complexity and poor scalability.

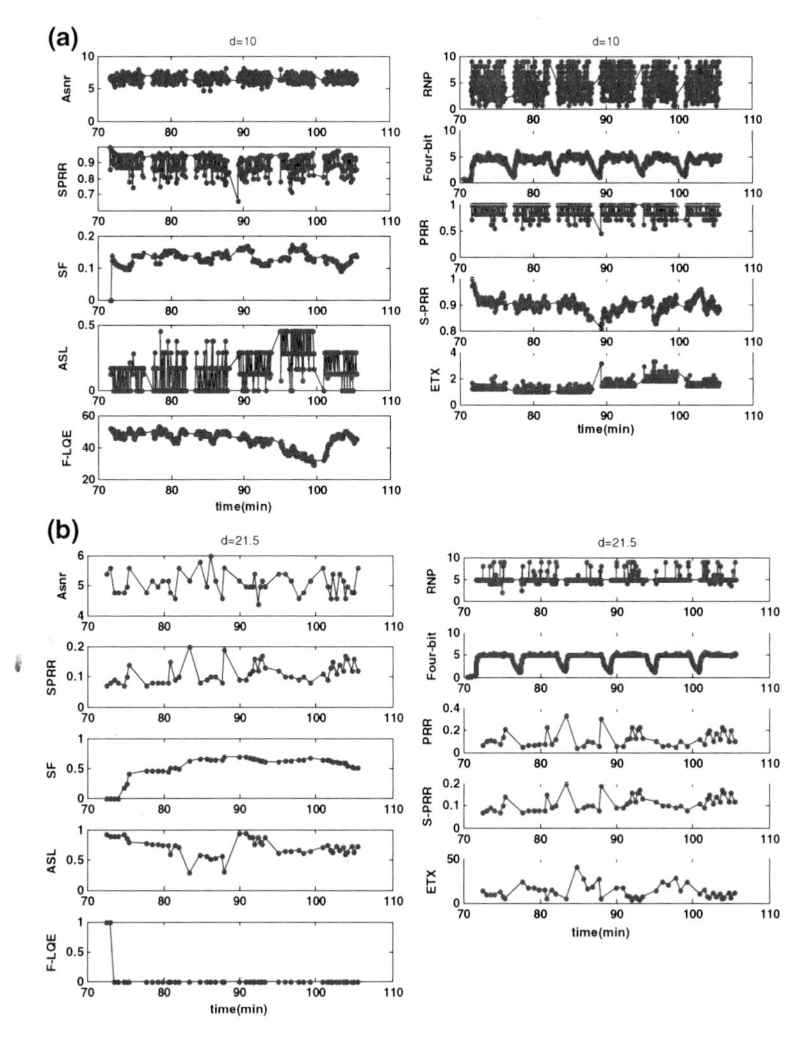

Fig. 4.13 Temporal behaviour of LQEs when faced with links with different qualities, observed by Tossim 2 simulation. **a** Moderate link. **b** Bad link

4.5.2.3 TOSSIM 2 Realism

In the above discussion, we have compared the simulation results to the experimental results. Hence, we can argue that TOSSIM 2 channel model provides a reasonable tradeoff between accuracy and simplicity. Nevertheless, recall that despite that TOSSIM 2 channel model captures important link properties, including spatial and temporal behaviors, and link asymmetry, it does not takes into account the variation of the RSS according to the direction. In addition, TOSSIM 2 channel model assumes a static environment. Consequently, the RSS is constant according to time.

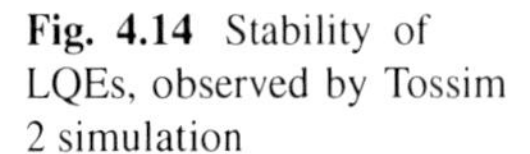

Fig. 4.14 Stability of LQEs, observed by Tossim 2 simulation

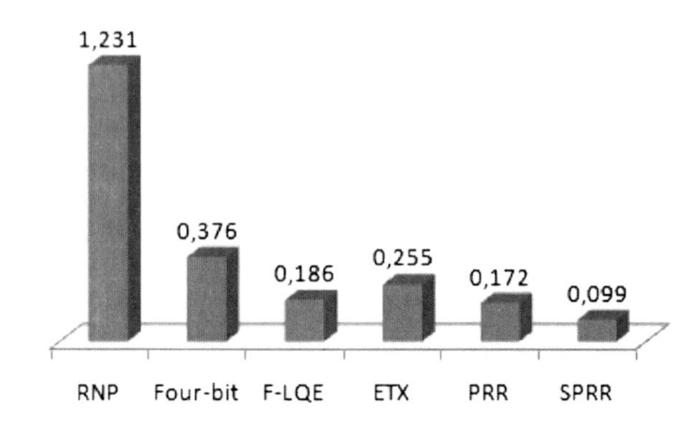

What makes the channel variability is only the noise floor variation. In our study, these simplifications had not a great impact on the validity of the results, but they could be for other link-layer dependant studies.

4.6 Lessons Learned

Lesson 1: We have drawn the following recommendations for the design of an efficient link quality estimator (LQE):

- An efficient LQE must be reactive to persistent (long-term) variations in link quality, yet stable by ignoring transient (short-term) variations in the link quality. A good balance between reactivity and stability can be obtained through the use of EWMA (Exponentially Weighted Moving Average) filtering [1]. This filter has two parameters that should be carefully tuned: the smoothing factor α and the estimation window w.
- Efficient link quality estimation that provides a fine-grained classification of links, especially intermediate quality links, should be based on several link quality metrics, each metric capturing a particular link property such as link asymmetry or stability. In fact, a single metric (e.g., RSSI, PRR, RNP, ETX) can only assess a particular link property and thus provides just a partial characterization of the link.
- A common error is to design a unidirectional LQE and use it for the assessment of bidirectional links, assuming that both link directions have the same quality, i.e., the link is symmetric. This misconception (typically found in mesh routing) has a negative impact on the performance of network protocols and can be detrimental to performance due to the prevalence of asymmetric links in low-power wireless networks. This does not mean that all LQEs should be bidirectional and unidirectional LQEs are useless. Specific applications such as data collection based on tree routing only need to assess one link direction (from child to parent). Thus, using a unidirectional LQE for these applications, specifically a sender-side LQE, is a convenient choice [16]. The design of bidirectional LQEs is not a trivial task.

A bidirectional LQE should combine feedback from both link directions. The main challenge is that these feedbacks should be obtained at the same time in order to cope with link dynamics. ETX and four-bit are bidirectional LQEs, but they do not meet this challenge.

Lesson 2: PRR and RNP are two representative and basic LQEs. They have been extensively used for routing protocols and also for the design of composite LQEs. Hence, it is important to understand their features:

- For good-quality and bad-quality links, i.e., links having high (e.g., >90 %) and low reception rates (e.g., <50 %) respectively, PRR follows the same behavior as RNP. However, for intermediate quality links, PRR *overestimates* the link quality because it does not take into account the underlying distribution of packet losses. When the link exhibits short periods during which packets are not received, the PRR can still have high value but the RNP is high so that it indicates the real link state. As a matter of fact, a packet that cannot be delivered may be retransmitted several times before aborting transmission.
- RNP is more reactive than PRR but it can underestimate link quality. In fact, RNP is a sender side LQE, i.e., it is computed based on transmitted packets. Consequently, RNP is able to provide link quality estimates as long as there is traffic generated from the sender. On the other hand, PRR is receiver side, i.e., it is computed based on received packets. Consequently, when the link is of poor quality, packets are not delivered and PRR cannot be computed. However, RNP can underestimate link quality in particular situations, as sometimes packets are retransmitted many times before being successfully received. This situation yields good PRR but bad RNP. Further, RNP is unstable compared with PRR.

Lesson 3: The design of LQEs that provide a holistic view on the link quality is a relatively new research problem, thus, several research challenges still remain open. One challenging problem is to select representative link quality metrics for the specification of a holistic link quality estimation. For instance, a big emphasis has been made in the literature about the goodness of hardware metrics namely RSSI, LQI, and SNR in quantifying some properties of the link. Another challenging problem is to devise convenient techniques for combining these metrics and producing a single link quality estimate. F-LQE addresses these challenges as it combines four link metrics (SPRR, ASNR, ASL, and SF) using Fuzzy Logic. However, it has the following limitations:

First, F-LQE is computationally more complex than traditional LQEs, such as PRR, RNP, four-bit, ETX, and WMEMWA. This is due to the fact that F-LQE computes four link quality metrics (SPRR, ASL, SF, and ASNR/ALQI), applies these metrics to piecewise linear membership functions, then combines the different membership levels into a given equation. In contrast, traditional LQEs are either based on a single link quality metric, or combine fewer metrics than F-LQE. For example, four-bit combines two link quality metrics through a simple weighted sum (the EWMA filter). F-LQE also consumes more memory, which is a consequence of its computation complexity (refer to Chap. 5).

Table 4.3 Comparison of considered LQEs

	Stability	Over-estimation	Under-estimation	Reactivity
ETX	☹	☹	-	☹
four-bit	😐	-	☹	☺
PRR	☺	☹	-	☹
RNP	☹	-	☹	☺
WMEWMA	☺	☹	-	☹
F-LQE	☺	-	-	☹

Second, F-LQE is a receiver-side LQE as all considered link quality metrics are computed based on received packets. Consequently, when the link is of poor quality, packets are not delivered and F-LQE estimates can not be updated. This limitation has negative impact on mobility management schemas, where responsiveness to link quality dynamics is a major concern. Nevertheless, F-LQE stands for the methodology of combing representative link quality metrics using Fuzzy Logic, for a holistic characterization of the link. Hence, another possible version of F-LQE would be the combination of sender-side link quality metrics such as RNP, in order to better cope with link dynamics.

4.7 Conclusion

We have thoroughly analyzed and compared several well-know LQEs, namely PRR, WMEWMA, ETX, RNP, four-bit, and F-LQE, by analyzing their statistical properties independently from higher layer protocols. The results of this study are summarized in Table 4.3. In next chapter, we investigate the impact of link quality estimation on higher layer protocols.

References

1. Woo A, Culler D (2003) Evaluation of efficient link reliability estimators for low-power wireless networks. Technical Report UCB/CSD-03-1270, EECS Department, University of California, Berkeley (2003). http://www.eecs.berkeley.edu/Pubs/TechRpts/2003/6239.html
2. Cerpa A, Wong JL, Potkonjak M, Estrin D (2005) Temporal properties of low power wireless links: modeling and implications on multi-hop routing. In: Proceedings of the 6th international symposium on mobile ad hoc networking and computing (MobiHoc '05). ACM, pp 414–425
3. Fonseca R, Gnawali O, Jamieson K, Levis P (2007) Four bit wireless link estimation. In: Proceedings of the 6th international workshop on hot topics in networks (HotNets VI). ACM SIGCOMM

4. Baccour N, Koubâa A, Jamaa MB, do Rosário D, Youssef H, Alves M, Becker LB (2011) Radiale: a framework for designing and assessing link quality estimators in wireless sensor networks. Ad Hoc Netw 9(7):1165–1185
5. Srinivasan K, Kazandjieva MA, Jain M, Kim E, Levis P (2008) Demo abstract: swat: enabling wireless network measurements. In: Proceedings of the 6th ACM conference on embedded network sensor systems (SenSys '08). ACM, pp 395–396
6. Gomez C, Boix A, Paradells J (2010) Impact of LQI-based routing metrics on the performance of a one-to-one routing protocol for IEEE 802.15.4 multihop networks. EURASIP J Wirel Commun Netw 2010:6:1–6:20
7. Rondinone M, Ansari J, Riihijärvi J, Mähönen P (2008) Designing a reliable and stable link quality metric for wireless sensor networks. In: Proceedings of the workshop on real-world wireless sensor networks (RealWSN). Glasgow, Scotland, pp 6–10
8. Boano CA, Zúñiga MA, Voigt T, Willig A, Römer K (2010) The triangle metric: fast link quality estimation for mobile wireless sensor networks. In: Proceedings of the 19th international conference on computer communications and networks (ICCCN), pp 1–7
9. Levis P, Lee N, Welsh M, Culler D (2003) Tossim: accurate and scalable simulation of entire tinyos applications. In: Proceedings of the 1st international conference on embedded networked sensor systems (SenSys '03). ACM, pp 126–137
10. TinyOS. http://www.tinyos.net/
11. Gay D, Levis P, von Behren R, Welsh M, Brewer E, Culler D (2003) The nesc language: a holistic approach to networked embedded systems. In: Proceedings of the ACM SIGPLAN conference on programming language design and implementation (PLDI '03). ACM, pp 1–11
12. Lee H, Cerpa A, Levis P (2007) Improving wireless simulation through noise modeling. In: Proceedings of the 6th international conference on information processing in sensor networks (IPSN '07). ACM, pp 21–30
13. Zuniga M, Krishnamachari B (2007) An analysis of unreliability and asymmetry in low-power wireless links. ACM Trans Sen Netw 3(2):63–81
14. Rappapport TS (2001) Wireless communications: principles and practice. Prentice Hall, Upper Saddle River
15. Topology configuration (2009). http://www.tinyos.net/dist-2.0.0/tinyos-2.x/doc/html/tutorial/usc-topologies.html
16. Baccour N, Koubaa A, Ben Jamaa M, Youssef H, Zuniga M, Alves M (2009) A comparative simulation study of link quality estimators in wireless sensor networks. In: Proceedings of the 17th IEEE/ACM international symposium on modelling, analysis and simulation of computer and telecommunication systems (MASCOTS '09). IEEE, pp 301–310
17. Werner-Allen G, Swieskowski P, Welsh M (2005) Motelab: a wireless sensor network testbed. In: Proceedings of the int symposium on information processing in sensor networks (IPSN '05). IEEE Press, pp 483–488
18. Chun BN, Buonadonna P, AuYoung A, Ng C, Parkes DC, Shneidman J, Snoeren AC, Vahdat A (2005) Mirage: a microeconomic resource allocation system for sensornet testbeds. In: IEEE workshop on embedded networked sensors. IEEE Computer Society, pp 19–28
19. Handziski V, Köpke A, Willig A, Wolisz A (2006) Twist: a scalable and reconfigurable testbed for wireless indoor experiments with sensor networks. In: Proceedings of the 2nd international workshop on multi-hop Ad hoc networks: from theory to reality (REALMAN '06). ACM, pp 63–70
20. Arora A, Ertin E, Ramnath R, Nesterenko M, Leal W (2006) Kansei: a high-fidelity sensing testbed. IEEE Internet Comput 10(2):35–47. doi:10.1109/MIC.2006.37
21. Johnson D, Flickinger D, Stack T, Ricci R, Stoller L, Fish R, Webb K, Minor M, Lepreau J (2005) Emulab's wireless sensor net testbed: true mobility, location precision, and remote access. In: Proceedings of the 3rd international conference on embedded networked sensor systems (SenSys '05). ACM, pp 306–306. http://doi.acm.org/10.1145/1098918.1098971
22. Cerpa A, Busek N, Estrin D (2003) Scale: a tool for simple connectivity assessment in lossy environments. Technical report

23. Srinivasan K, Levis P (2006) Rssi is under appreciated. In: Proceedings of the 3th workshop on embedded networked sensors (EmNets)
24. Radiale benchmarking tool (2010). http://www.open-LQE.net
25. Tinyos 2 (2009). http://www.tinyos.net/tinyos-2.x/tos/
26. Polastre J, Szewczyk R, Culler D (2005) Telos: enabling ultra-low power wireless research. In: Proceedings of the 4th internatioanl symposium on information processing in sensor networks (IPSN '05). IEEE Press, pp 364–369
27. Chipcon cc2420: Data sheet (2009). http://enaweb.eng.yale.edu/drupal/system/files/CC2420_Data_Sheet_1_4.pdf
28. Zuniga M, Krishnamachari B (2004) Analyzing the transitional region in low power wireless links. In: Proceedings of the 1st international conference on sensor and Ad hoc communications and networks (SECON '04). IEEE Communications Society, pp 517–526

Chapter 5
On the Use of Link Quality Estimation for Improving Higher Layer Protocols and Mechanisms

Abstract Several network protocols and mechanisms rely on efficient link quality estimation to mitigate the unreliability of low-power links. A LQE can be efficient on a per-link basis, but it may lead to a poor performance when integrated in a particular protocol or mechanism. The objective of this chapter is to show how to use link quality estimation for improving higher layer protocols and mechanisms—especially routing and mobility management.

5.1 Introduction

Link quality estimation plays a crucial role for both routing protocols and mobility management mechanisms.

The consideration of link quality in the process of routing is a prerequisite to overcome link unreliability and maintain acceptable network performance. Indeed, delivering data over high quality links (i) reduces the number of packet retransmissions in the network, (ii) increases its throughput and (iii) ensures a stable topology. This implies that efficient routing metrics should integrate not only the path length criterion, in terms of hops or communication delay, but also the path's *global quality*. Path quality is evaluated based on the assessment of links that compose it. Hence, in the first part of this chapter, we address the question of how to design an efficient routing metric based on a novel LQE that ensures reliable end-to-end delivery.

Mobility management is a wide area that covers various aspects such as handoff processes, re-routing, re-addressing and security issues. Within the scope of this book, our main focus is on the *handoff processes* as it leverages on reliable link quality estimation. Handoff refers to the process where a mobile node disconnects from one point of attachment and connects to another. Hence, it is clear that handoff process greatly relies on link quality estimation. This fact motivates us to address the question of how to use link quality estimation for an efficient handoff in mobility management solutions.

N. Baccour et al., *Radio Link Quality Estimation in Low-Power Wireless Networks*, SpringerBriefs in Electrical and Computer Engineering,
DOI: 10.1007/978-3-319-00774-8_5,

5.2 On the Use of Link Quality Estimation for Routing

5.2.1 Link Quality Based Routing Metrics

Link quality-based routing metrics consider the criteria of path global quality in path selection. They may also integrate other criteria such as path length in terms of hop count or communication delay, and nodes energy, depending on the application requirements.

Path quality is determined through the assessment of links composing the path. Depending on the link quality estimator category, the path quality can be the sum (e.g., for RNP-based LQEs), the product (e.g., PRR-based LQEs), the max/min (e.g., Hardware-based LQEs and score-based LQEs), or any other function, of link quality estimates over the path. Next we overview a set of representative routing metrics to illustrate this statement.

The DoUble Cost Field HYbrid (DUCHY) [1] and SP(t) [2] are two routing metrics that select routes with minimum hops and high quality links. For DUCHY, each node maintains a set of neighbors that are nearer, in terms of hops, to the tree root. Then, the parent node is selected among the maintained set of neighbors as the one that has the best link quality. Link quality estimation is performed using both CSI (Channel State Information) and RNP. As for SP(t), each node maintains a set of neighbors that have link quality exceeding a threshold t. Link quality estimation is performed using WMEWMA. Then, the parent node is selected among the maintained set of neighbors as the nearest one, in terms of hops, to the tree root.

ETX [3] and four-bit [4] are two link quality estimators that have been extensively used as routing metrics. Both approximate the RNP (RNP-based category). Using ETX or four-bit, the path cost is the sum of quality estimates of its links. This path cost can be generalized to any RNP-based link estimator, since the number of packet retransmissions over the path is typically the sum of packet retransmissions of each link composing the path.

MAX-LQI and Path-DR [5] aim to select the most reliable path, regardless of its hop count. MAX-LQI selects the path having the highest minimum LQI over the links that compose the path. Path-DR approximates the link PRR using LQI measurements and then evaluates the path cost as the product of link PRRs. Path-DR selects paths having the maximum of this product. The product of link estimates can be generalized to any PRR-based LQE.

The aforementioned link quality based routing metrics use traditional LQEs, such as PRR, RNP, four-bit, and LQI. These LQEs are not sufficiently accurate as they either rely on a single-link-quality metric, or use simple but inaccurate techniques for the combination of link quality metrics such as filtering through the EWMA. Further, these metrics can only capture one link aspect such as link delivery or the number of packet retransmissions over the link (refer to Chap. 4 for more details on the limitation of these LQEs). On the other hand, F-LQE was shown more reliable and more stable than these LQEs as it takes into account several important link aspects. The accuracy of link quality estimation greatly affects the effectiveness of link quality based routing metrics.

In [6], the authors propose using F-LQE to design an efficient link quality based routing metric. In other words, the authors addressed the question of how to use reliable link quality estimation provided by F-LQE to build a routing metric that improves routing performance, e.g., in terms of end-to-end packet delivery. They call their routing metric FLQE-RM (Fuzzy Link Quality Estimator based Routing Metric). FLQE-RM has three main design requirements:

- First, FLQE-RM should correctly evaluate the path cost based on individual link costs, i.e., F-LQE link quality estimates. This requirement should be carefully addressed as F-LQE can be efficient on a link basis, but inadequate at the path level due to inadequate path cost evaluation. This situation may result in a dramatic reduction of routing performance.
- Second, path cost evaluation should take into account not only the path global quality but also the weakest quality link in the path. In fact, a path may have the highest global quality among candidate paths; yet it may still contain a weak quality link. This situation leads to several packet looses over this link, which negatively affects the routing performance, such as the end-to-end packet delivery.
- Third, FLQE-RM should favor the selection of short paths. In fact, selecting short paths reduces the number of transmissions over the path and also the number of nodes involved for packet delivery, which conserves the node's energy and thus extends the network lifetime.

Based on these requirements, FLQE-RM is defined as follow:

$$FLQE\_RM = \sum_{i \in Path} \frac{1}{FLQE_i} \tag{5.1}$$

$\frac{1}{FLQE_i}$ is the cost of the link i. Thus FLQE-RM defines the path cost as the sum of the links' costs. The path having minimal cost is selected. FLQE-RM takes into account the global path quality and implicitly favors the selection of short paths thanks to the link cost definition. Indeed, by defining the link cost as $\frac{1}{FLQE_i}$ instead of $FLQE_i$, the path selection is a minimization of the path cost instead of a maximization. Hence, the longer the path, higher its cost and thus the lower the chance it will be selected. The link cost definition also improves the effectiveness of FLQE-RM by avoiding paths having weak quality links: The lower the link quality, the higher its cost, which impacts the overall path cost and increases the probability that the path is rejected.

Next, we show how FLQE-RM can indeed improve the routing performance, when integrated to the CTP (Collection Tree Protocol) routing protocol.

5.2.2 *Overview of CTP (Collection Tree Protocol)*

As data collection is one of the most popular low-power wireless networks applications, CTP has gained a lot of interest during the last years. CTP establishes and maintains a routing tree, where the tree root is the ultimate sink node of the collected data. In CTP, three types of nodes can be identified:

- *The sink node*: One node in the network advertises itself as a sink node (generally the node with id 0). It is the root of the routing tree. All other nodes forward information to the root based on the tree formed via link quality estimation.
- *The parent node*: Except the sink node, each node has a parent, which represents the next hop towards the tree root. Each parent node has a certain number of child nodes.
- *The child node*: It is associated to a single parent node and can be in turn the parent of other child nodes situated further below in the tree hierarchy. Notice that the data traffic flow is from the child node to the parent node.

CTP is the reference protocol for the network layer of TinyOS 2.x. stack [7]. Due to its modularity, and also to the fact that it relies on a link quality based routing metric, we use it as a benchmark for analyzing the impact of different link quality based routing metrics on routing performance.

The CTP implementation contains three basic components: the link estimator, the routing engine and the forwarding engine. These components are shortly described next.

5.2.2.1 Link Estimator

This component is based on the Link Estimation Exchange Protocol (LEEP) [8] and four-bit [4]. Note that the implementation of four-bit in Link estimator component is slightly different from its specification in [4]. According to this implementation, four-bit combines beacon-driven estimate (estETX) and data-driven estimate (RNP) using the EWMA filter. RNP is computed based on *DLQ* transmitted/retransmitted data packets and estETX is computed based on *BLQ* received beacons. It is given by the following expression:

$$ETX(BLQ, \alpha) = \frac{1}{SPRR_{in} \times SPRR_{out}} - 1 \tag{5.2}$$

where $SPRR_{in}$ is the PRR of the inbound link, smoothed using EWMA. $SPRR_{out}$ is the PRR of the outbound link, smoothed using EWMA. $SPRR_{out}$ is gathered from a received beacon or data packet.

Each node maintains a *neighbor table*, where each entry contains useful information for estimating the quality of the link to a particular neighbor. These information include (i) the neighbor address; (ii) the sequence number of the last received beacon, the number of received beacons and the number of missed beacons (these are used for $SPRR_{in}$ computation); (iii) the inbound link quality ($SPRR_{in}$) and the outbound link quality ($SPRR_{out}$); (iv) the number of acknowledged packets and the total number of transmitted/retransmitted data packets (these are used to compute $estETX_{up}$); (v) link cost (four-bit estimate); and (vi) different flags that describe the state of the entry.

The replacement policy in the neighbor table is governed by the use of the *compare bit* and the *pin bit*. The pin bit applies to the neighbor table entries. When the pin bit is set on a particular entry, it cannot be removed from the table until the pin bit is cleared. The compare bit is checked when a beacon is received. It indicates whether the route provided by the beacon sender is better than the route provided by one or more of the entries in the neighbor table.

5.2.2.2 Routing Engine

This component is responsible for the establishment and maintenance of the routing tree for data collection. Each node maintains a *routing table*. In this table, there is an entry for each where each neighbor. An entry contains .the following fields:

- the address of the neighbor,
- the address of the parent of this neighbor,
- the cost of the neighbor, and
- an indicator on whether the neighbor is congested.

The neighbor cost refers to the route cost from this neighbor to the sink. Generally, a node cost is computed as the cost of its parent plus the cost of its link to its parent. The link cost corresponds to four-bit estimate.The cost of a route is computed as the sum of links' costs. Lower route costs are better. Note that the sink node has a cost equal to zero.

A node updates its route to the sink, which corresponds to the update of its parent, periodically. Parent update consists in searching in the routing table for a neighbor that provides a route cost better than that provided by current parent. To compute the route cost through a given neighbor, the node gets the neighbor cost from the routing table, and the link cost to the neighbor from the neighbor table and then sums the two values. To avoid frequent parent changes leading to unstable topology, a node changes its parent only when a number of conditions are satisfied. For example, the new parent should provide a route cost lower than the current route cost by *ParentChTh*, which is a constant parameter defined by CTP.

The tree is maintained by beacons sent by each node according to an adaptive beaconing rate, to ensure a minimum number of beacons sent along with a consistent tree. When a node sends a beacon, it includes the address of its parent as well as its cost, i.e., the route cost from the node to the sink, in the beacon header. It also includes a list of *neighbor entries* in the beacon footer. A neighbor entry is composed of the neighbor address and the SPRR of the inbound link, $SPRR_{in}$. When a node receives a beacon, it seeks for its address in the list of neighbor entries. When found, it extracts the $SPRR_{in}$, and updates the SPRR of the outbound link, $SPRR_{out}$, in its neighbor table.

5.2.2.3 Forwarding Engine

This component is responsible for queueing and scheduling outgoing data packets. Each node, maintains a forwarding queue that adopts a set of rules to process data packets. For example, a data packet is ejected from the queue if it has been acknowledged or has reached the maximum retransmission count. When a node receives a data packet from a neighbor with cost less than its cost, it drops the packet and signals an inconsistency in the network (a loop detection). Data packets are automatically forwarded to the next hop in the tree, which corresponds to the parent node. When a node sends a data packet, it includes its cost in the packet header. As with beacons, the node includes a list of *neighbor entries* in the packet footer.

5.2.3 Integration of FLQE-RM in CTP

To integrate the proposed F-LQE based routing metrics in CTP, we have implemented F-LQE in the Link Estimator component, as replacement of the four-bit estimator.

Beacon-Driven Link Quality Estimation

Recall that F-LQE combines four metrics, which are computed at the receiver side, i.e., based on received traffic:

- SPRR: Smoothed Packet Reception Ratio over the link,
- ASL: link ASymmetry Level,
- SF: link Stability Factor, and
- ASNR: link Average Signal-to-Noise Ratio.

Our implementation of F-LQE leverages on broadcast control traffic (i.e., beacons), which is initiated by the CTP routing engine for the topology control. F-LQE can be also implemented based on data traffic, which requires the overhearing of incoming packets.

CTP uses an adaptive beaconing rate that changes according to the topology consistency. In our implementation, we disabled this mechanism and we used a constant beaconing rate of 1 beacon/s.

Channel Quality Assessment

In F-LQE, channel quality is assessed by ASNR. However, SNR is not the optimum choice in the context of routing, where the node needs to quickly switch to a better parent when the current parent breaks down. In fact, SNR computation is relatively time consuming as it involves two separated operations: It is derived by subtracting the noise floor (N) from the received signal (S), where the S is deduced by sampling the RSSI at the packet reception, and N is derived from the RSSI sample just after the packet reception. Therefore, in F-LQE implementation, we substitute SNR by LQI (Link Quality Indicator), which assesses channel quality in one operation while still providing acceptable accuracy.

Link Direction

In CTP tree routing, data travel from child to parent. In order to select their parents, child nodes need to assess direct links, i.e., $child \rightarrow parent$ links. Although F-LQE takes into consideration link asymmetry through ASL metric, it evaluates the reverse link, i.e., $parent \rightarrow child$ link (because each of SPRR, SF, and ASNR provides reverse link estimate). Considering the reverse link estimate to decide about the direct link for parent selection leads to misleading routing decisions. Therefore, we define two F-LQE estimates: $\mathrm{F-LQE}_{in}$ and $\mathrm{F-LQE}_{out}$. $\mathrm{F-LQE}_{in}$ is the F-LQE for the reverse link, (i.e., inbound link). It is computed by each node, based on incoming beacons. $\mathrm{F-LQE}_{out}$ is the F-LQE for the direct link, (i.e., outbound link) and it is gathered from received packets. $\mathrm{F-LQE}_{in}$ and $\mathrm{F-LQE}_{out}$ are stored in the neighbor table, with respect to each neighbor node. As reported in Sect. 5.2.2, CTP defines a list of neighbor entries, that is included in the footer of each sent packet. In our implementation, a neighbor entry is composed of the neighbor address, PRR_{in}, and $\mathrm{F-LQE}_{in}$. When a node receives a packet, it extracts PRR_{in}, and $\mathrm{F-LQE}_{in}$ and stores them in its neighbor table, specifically in PRR_{out}, and the $\mathrm{F-LQE}_{out}$ fields. Link Estimator component is used by the Routing engine to get the link cost, which corresponds to $\frac{1}{F-LQE_{out}}$.

Parent Update

Nodes update their parents, when the new parent is better than the current one by *ParentChTh*. This constant parameter depends on the routing metric. We set it to 4 for F-LQE based routing metrics (based on several experimental measurements). The *ParentChTh* for four-bit is equal to 1.5 (default value).

Routing Engine

Like four-bit, FLQE-RM selects parents that lead to minimal path costs, where a path cost is the sum of its link costs. Hence, the implementation of FLQE-RM does not require major modifications in the Routing Engine component.

5.2.4 Impact of FLQE-RM on CTP Performance

In this section, we investigate the impact of FLQE-RM on the performance of CTP using experimentation with real WSN platforms. Further, we compare the impact of FLQE-RM to that of four-bit, the default metric of CTP, as well as ETX [3]. Both four-bit and ETX are considered by the community as representative and reference metrics.

In our study, the considered performance metrics are the following:

- Packet Delivery Ratio (PDR). It is computed as the total number of delivered packets (at the sink node, i.e., the root) over the total number of sent packets (by all source nodes). This metric indicates the end-to-end reliability of routing protocols.
- Average number of retransmissions across the network per delivered packet. (RTX). This metric is of paramount importance for low-power wireless networks as it greatly affects the network lifetime. In fact, communication is the most energy consuming operation for a sensor node. Therefore, efficient routing protocols try to minimize packet retransmissions by delivering data over high quality links, which extend the network lifetime.
- Average number of parent changes per node (ParentCh). This metric is an indicator of topology stability. The number of parent changes depends on two factors: the *ParentChTh* parameter of CTP, and also the agility of the LQE allowing for detecting link quality changes. Too many parent changes leads to instable topology, but improves the quality of routes and thus improves routing performance (e.g., PDR and RTX). On the other hand, few parent changes leads to stable topology but also paths with potentially lower quality. Hence, an agile LQE, along with a good *ParentChTh* choice would lead to a good tradeoff between topology stability and route quality.
- Average path lengths, i.e., average Hop Count. It is important that link quality aware routing protocols minimize route lengths in order to reduce (i) the number of packet transmissions to deliver a packet, (ii) the number of involved nodes for data delivery, and possibly (iii) the end-to-end latency (in case the involved nodes are not overloaded).

5.2.4.1 Experiments Description

In our experimental study, we resort to remote testbeds (i.e., general-purpose testbeds) for large scale experiments. Examples of remote testbeds include MoteLab [9], Indriya [10], Twist [11], Kansei [12], and Emulab [13].

Remote testbeds are designed to be remotely used by several users over the world. Roughly, they are composed of four building blocks: (i) the underlying low-power wireless network (i.e., a set of sensor nodes), (ii) a network backbone providing reliable channels to remotely control sensor nodes, (iii) a server that handles sensor nodes reprogramming and data logging into a database, and (iv) a web-interface coupled with a scheduling policy to allow the testbed sharing among several users. The testbed users must be experts on the programming environment supported by the tesbeds (e.g. TinyOS, Emstar), to be able to provide executable files for motes programming. They must also create their own software tool to analyze the experimental data and produce results.

Our experimental study is carried out on both MoteLab [9] and Indriya [10] testbeds. MoteLab consists of 190 TMote Sky motes, deployed over three floors of Harvard university building, and Indriya consists of 127 TelosB motes, deployed over three floors in the National University of Singapore (NUS) building. In both testbeds, node placement is very irregular. Node programming is performed using TinyOS.

In contrast to Indriya, which is a recently released testbed, MoteLab is serving the WSN community for six years. Hence, around 100 nodes in MoteLab are not working mostly due to aged hardware. Further, the number of working nodes in both testbeds varies according to time due to many reasons such as hardware failure and human activity. Our experiments were conducted within April–July 2011, where 72 nodes from Motelab and 121 nodes from Indriya were available.

Using low transmission powers for sensor nodes leads to more intermediate quality links, and thus allows us better evaluate link quality based routing metrics. However, this may lead to a partitioned network, as some nodes may not be able to join the network due to poor connectivity. Hence, the transmission power should be correctly set to have as much as possible a rich set of links (i.e., having different qualities), while preserving the network connectivity. To this end, we set the transmission power to −25 dBm for Indriya experiments and to 0 dBm for MoteLab experiments. These values were determined through several experiments. In each experiment, we set the transmission power to different values and check the connectivity of the network through the graphical interface provided by the testbed software.

Our experiments consist of a many-to-one application scenario where nodes generate traffic at a fixed rate, destined to the sink node. Data collection is performed using CTP, with a fixed beacon rate (1 packet(pkt)/s). Nodes use the default MAC protocol in TinyOS, B-MAC. Recall that we set the transmit power to −25 dBm for Indriya experiments and to 0 dBm for MoteLab experiments. The radio channel is set to 26 to avoid interference with co-existing networks such as Wi-Fi. Most of experiments were conducted with Indriya as it provides more active nodes (121 nodes) than MoteLab (72 nodes). Each experiment lasts 60 min. Nodes begin their transmission after a delay of 10 min to enable the topology establishment.

Experiments are divided into different sets. In each experiment set, we varied a certain parameter to study its impact, and the experiment was repeated for each parameter modification. Parameters under-consideration were the testbed under use, traffic load, topology, and the number of source nodes. Table 5.1 depicts the different settings for each experiments set.

Table 5.1 Experiment sets

	Testbed	Traffic load (in pkts/s)	Number of source nodes	Topology (root ID)
Set 1: Impact of testbed	{Indriya, MoteLab}	0.125	120	1
Set 2: Impact of traffic load	Indriya	{0.125, 0.25, 0.5, 1, 2}	120	1
Set 3: Impact of source nodes	Indriya	0.125	{120, 84, 60, 42, 29}	1
Set 4: Impact of topology	Indriya	0.125	120	{1, 15, 58, 113}

5.2.4.2 Experimental Results

Performance for Different Testbeds

We begin by assessing the overall impact of FLQE-RM, four-bit and ETX on CTP routing performance, using the Indriya testbed (refer to Table 5.1—Set 1 of experiments). Each experiment is repeated 5 times. Experimental results are illustrated in Fig. 5.1.

Figure 5.1 shows that FLQE-RM provides better routing performance, compared to four-bit and ETX as it is capable to deliver more packets (Fig. 5.1a), with less retransmissions (Fig 5.1b), less parent changes (Fig. 5.1d), and through shorter routes (Fig. 5.1c).

Figure 5.1a shows that ETX has very low PDR compared with FLQE-RM and four-bit. This can be interpreted by the fact that ETX is not able to identify high quality routes for data delivery. One of the reasons is the unreliability of ETX as a LQE, i.e., ETX is not an accurate metric for link quality estimation. Further, ETX is unstable as it leads to frequent parent changes (Fig. 5.1d). Parent changes may lead to several packet looses. The unreliability and unstability of ETX was confirmed in Chap. 2, when we analyzed the statistical properties of different LQEs, including ETX, independently of higher layer protocols, especially routing.

Network conditions, especially the nature of the surrounding environment (e.g., indoor/outdoor, static/mobile obstacles, the geography of the environment), the type of the platform, and even the climate conditions (e.g., temperature, humidity), affect the quality of the underlying links, and thus impact the network performance. For this reason, we have investigated the performance of FLQE-RM, four-bit, and ETX, using a different testbed from Indriya. Experimental results carried out with MoteLab (refer to Table 5.1—Set 1 of experiments) are depicted in Fig. 5.2. From this figure two main observations can be made: First, by examining the PDR in Fig. 5.2a, it can be inferred that links in MoteLab have worse quality,than those in Indriya, as the maximum achieved PDR (by FLQE-RM) is equal to 75 %. Second, MoteLab experimental results confirm that FLQE-RM leads to the best routing performance

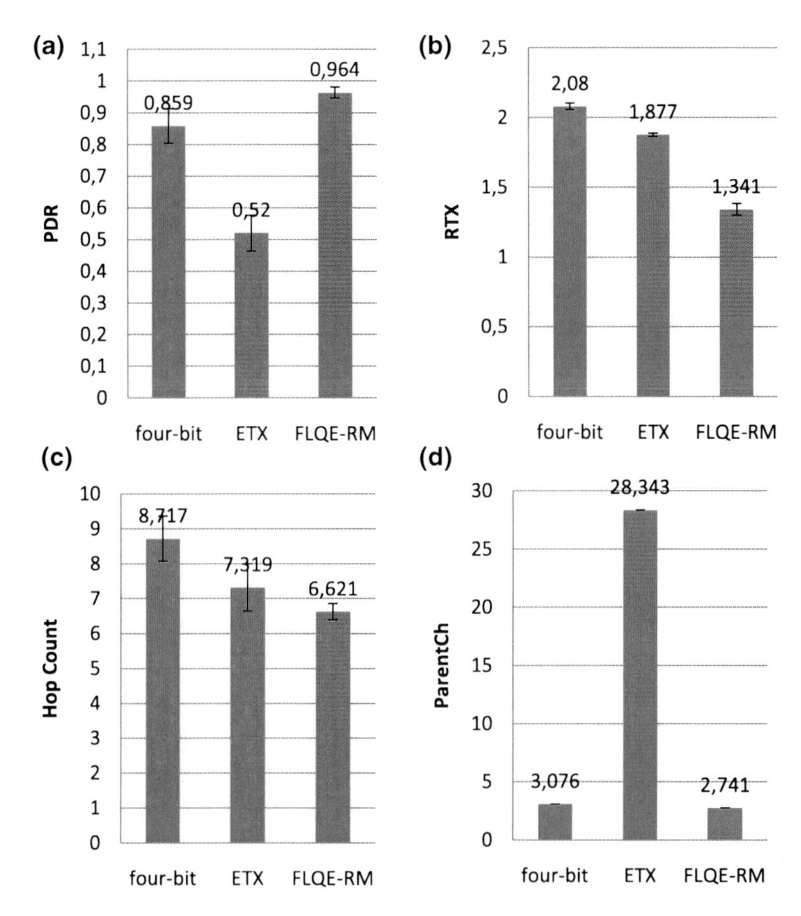

Fig. 5.1 **a** Packet delivery ratio (PDR). **b** Average umber of packet retransmissions (RTX). **c** Average routes hop count (Hop Count). **d** Average number of parent changes (ParentCh). Impact of FLQE-RM, four-bit, and ETX on CTP performance, using Indriya testbed (refer to Table 5.1—Set 1)

and ETX leads to the worst. This observation can be interpreted by F-LQE reliability. Indeed, we have shown in Chap. 3 that F-LQE provides a fine grain classification of links, especially intermediate links (better than four-bit and ETX).

Performance as a Function of the Traffic Load

We have assessed the impact of FLQE-RM, four-bit and ETX on CTP routing performance for different traffic loads. The Experiment settings are presented in Table 5.1—Set 2 and Fig. 5.3 illustrates the experimental results. With a higher traffic load, the

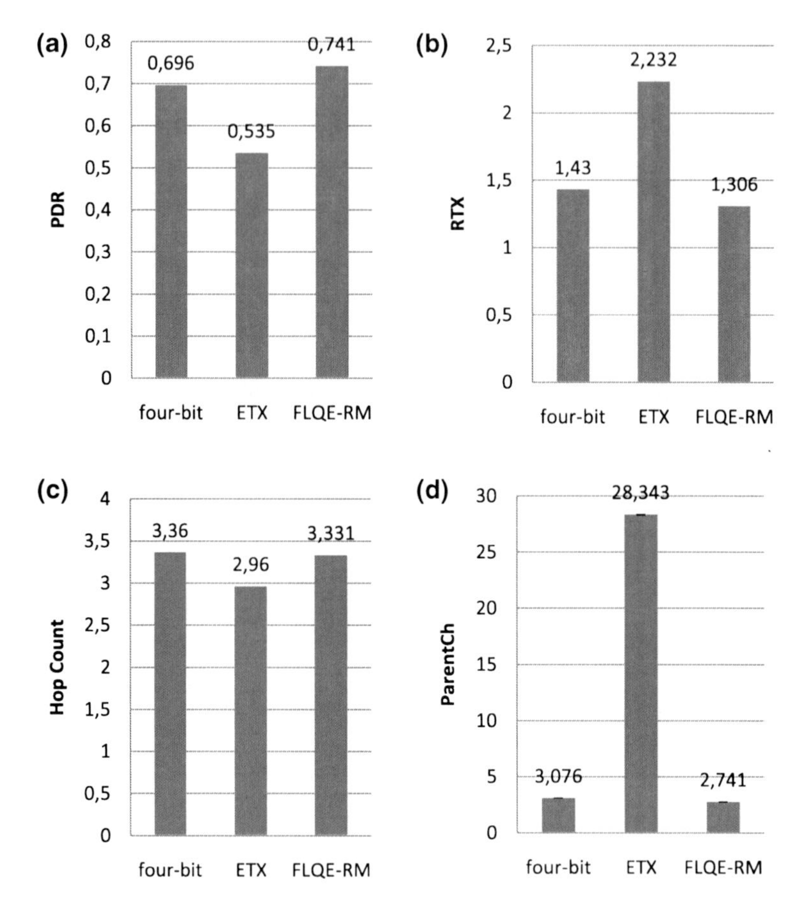

Fig. 5.2 **a** Packet delivery ratio (PDR). **b** Average number of packet retransmissions (RTX). **c** Average routes hop count (Hop Count). **d** Average number of parent changes (ParentCh2). Impact of FLQE-RM, four-bit and ETX, on CTP performance, using MoteLab testbed (refer to Table 5.1—Set 1)

congestion level of the network increases, which leads to packet losses induced by buffer overflows as well as MAC collisions.

For traffic loads less than or equal to 1 pkt/s, Fig. 5.3 shows that FLQE-RM performs better than four-bit and ETX: It increases the PDR and reduces the number of parent changes. If we observe RTX and Hop count together, it can be inferred that FLQE-RM reduces the global number of packet transmissions (i.e., Hop count) and retransmissions (i.e., RTX), compared with ETX and four-bit. For example, for traffic load equal to 1 pkt/s, FLQE-RM has RTX equal to 1.27 and Hop count equal to 4.56, while ETX has RTX equal to 1.123 and a Hop count equal to 4.86. Thus, overall, FLQE-RM reduces the number of packet transmissions and retransmissions (5.83) compared with ETX (5.98).

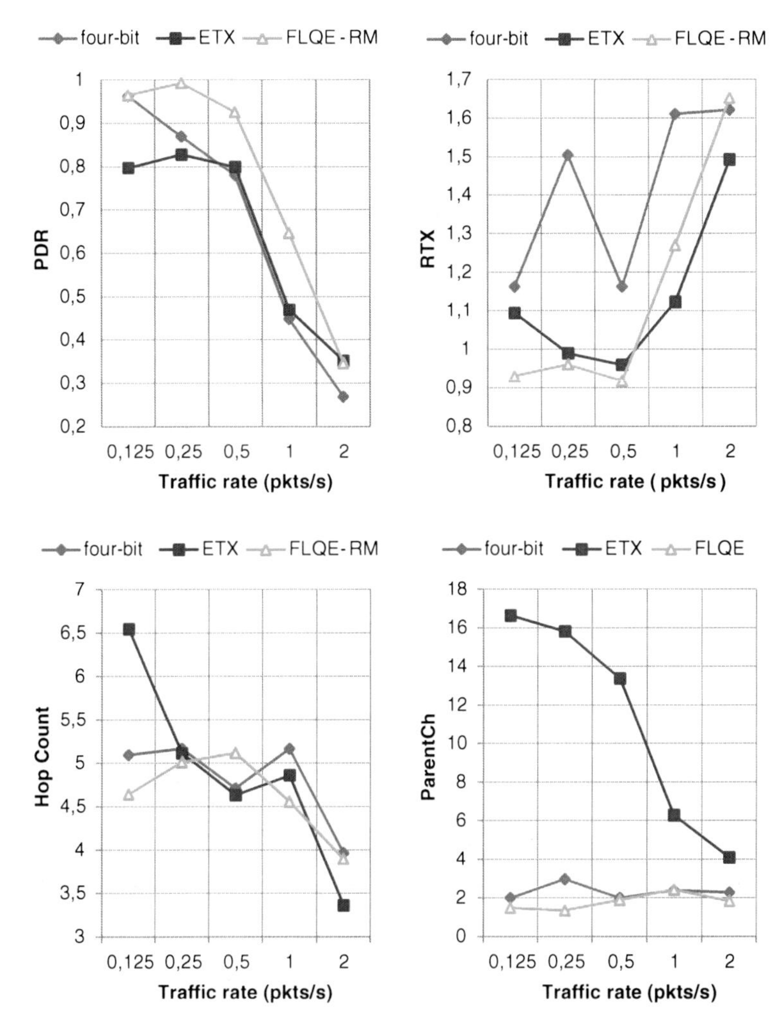

Fig. 5.3 Performance as a function of the traffic load (refer to Table 5.1—Set 2)

For traffic loads equal to 2 pkts/s, Fig. 5.3 shows that FLQE-RM provides slightly better (or nearly equal) performance than four-bit. This might be due to the fact that four-bit has more information on links status as the data rate (2 pkts/s) is double the beacon rate (1 pkt/s). Recall that four-bit uses both beacon traffic and data traffic for link quality estimation, while FLQE-RM and ETX perform link quality estimation based on beacon traffic only. Figure 5.3 also shows that for traffic loads equal to 2 pkts/s, ETX outperforms FLQE-RM and four-bit in terms of all performance metrics, except the parent changes. This observation would pertain to CTP, which does not contain any explicit congestion control mechanism, as it is designed for low data-rate applications.

Performance as a Function of the Number of Source Nodes

We have analyzed the impact of FLQE-RM, four-bit and ETX on CTP routing performance while varying the number of source nodes. the experiment settings are presented in Table 5.1—Set 3 and experimental results are illustrated in Fig. 5.4. By default, all nodes except the root node (i.e., 121 nodes) are data sources (refer to Table 5.1). By decreasing the number of source nodes, the congestion level of the network deceases, which reduces the number of packet looses induced by collisions or buffer overflow.

Figure 5.4 shows that overall, FLQE-RM leads to the best performance and ETX leads to the worst. By observing Figs. 5.3 and 5.4, it can be observed that generally, in terms of PDR, routing metrics are more sensitive to the traffic load variation than the number of source nodes variation. This is due to the considered data traffic rate (0.125 pkt/s), which is low enough to avoid network congestion for any number of source nodes.

Performance as a Function of the Topology

The network topology has a significant impact on routing performance [14]. To examine the impact of the topology on CTP routing, we considered different sink (root node) placements. Hence, for each CTP version, based on a particular routing metric (FLQE-RM, four-bit or ETX), we carried out a set of experiments, while varying the sink node assignment, i.e., varying the Root ID (refer to Table 5.1—Set 4).

Figure 5.5 illustrates the routing performance with respect to each routing metric as a function of the root ID assignment. This figure confirms the impact of the topology on routing performance. Further, it shows that again, FLQE-RM leads to the best performance and ETX leads to the worst, for all considered sink assignments.

5.2.4.3 Results Review

This section provides a review of our experimental results with the 122-node Indriya testbed, as illustrated in Tables 5.2, 5.3, and 5.4. These tables show that overall, FLQE-RM improves the end to end packet delivery (PDR) by up to 16 % over four-bit (Table 5.2) and up to 24 % over ETX (Table 5.4). It also reduces the number of retransmissions per delivered packet by up to 32 % over four-bit and also ETX (Table 5.3). The Hop count metric can be interpreted by the average route lengths as well as the average number of packet transmissions to deliver a packet. FLQE-RM reduces the Hop count by up to 4 % over four-bit (Tables 5.3 and 5.4) and up to 45 % over ETX (Table 5.3). The ParentCh metric implies on the topology stability. FLQE-RM improves topology stability by up to 47 % over four-bit (Table 5.3) and up to 92 % over ETX (Table 5.4).

Table 5.2 Overall results for Indriya experiments, where 121 nodes are data sources and the node with ID equal to 1 is selected as root, averaged over all considered traffic loads

Performance indicator	Four-bit	ETX	FLQE-RM
Packet delivery ratio (PDR)	0.666 ± 0.26	0.647 ± 0.19	0.775 ± 0.24
Number of retransmissions for packet delivery (RTX)	1.412 ± 0.2	1.131 ± 0.19	1.146 ± 0.28
Hop count	4.821 ± 0.45	4.902 ± 1.01	4.646 ± 0.43
Number of parent changes (ParentCh)	2.326 ± 0.35	11.246 ± 5.1	1.794 ± 0.37

Table 5.3 Overall results for Indriya experiments, where the traffic load is fixed to 0.125 pkt/s and the node with ID equal to 1 is selected as root, averaged over all considered number of source nodes

Performance indicator	Four-bit	ETX	FLQE-RM
Packet delivery ratio (PDR)	0.975 ± 0.02	0.822 ± 0.05	0.99 ± 0.01
Number of retransmissions for packet delivery (RTX)	1.059 ± 0.16	1.065 ± 0.13	0.719 ± 0.11
Hop count	5.613 ± 0.31	9.8 ± 2.13	5.402 ± 0.52
Number of parent changes (ParentCh)	2.585 ± 1.03	15.374 ± 2.05	1.363 ± 0.09

Table 5.4 Overall results for Indriya experiments, where 121 nodes are data sources and the traffic load is fixed to 0.125 pkt/s, averaged over all considered Root ID assignments

Performance indicator	Four-bit	ETX	FLQE-RM
Packet delivery ratio (PDR)	0.982 ± 0.01	0.793 ± 0.03	0.987 ± 0.01
Number of retransmissions for packet delivery (RTX)	0.991 ± 0.13	0.954 ± 0.09	0.769 ± 0.13
Hop count	5.268 ± 0.24	6.963 ± 0.48	5.129 ± 0.8
Number of parent changes (ParentCh)	1.247 ± 0.46	16.083 ± 2.1	1.354 ± 0.19

5.2.4.4 Memory Footprint and Computation Complexity

We measured the memory footprint with respect to four-bit, ETX, and FLQE-RM, in terms of RAM and ROM consumptions. As shown in Table 5.5, a sensor node

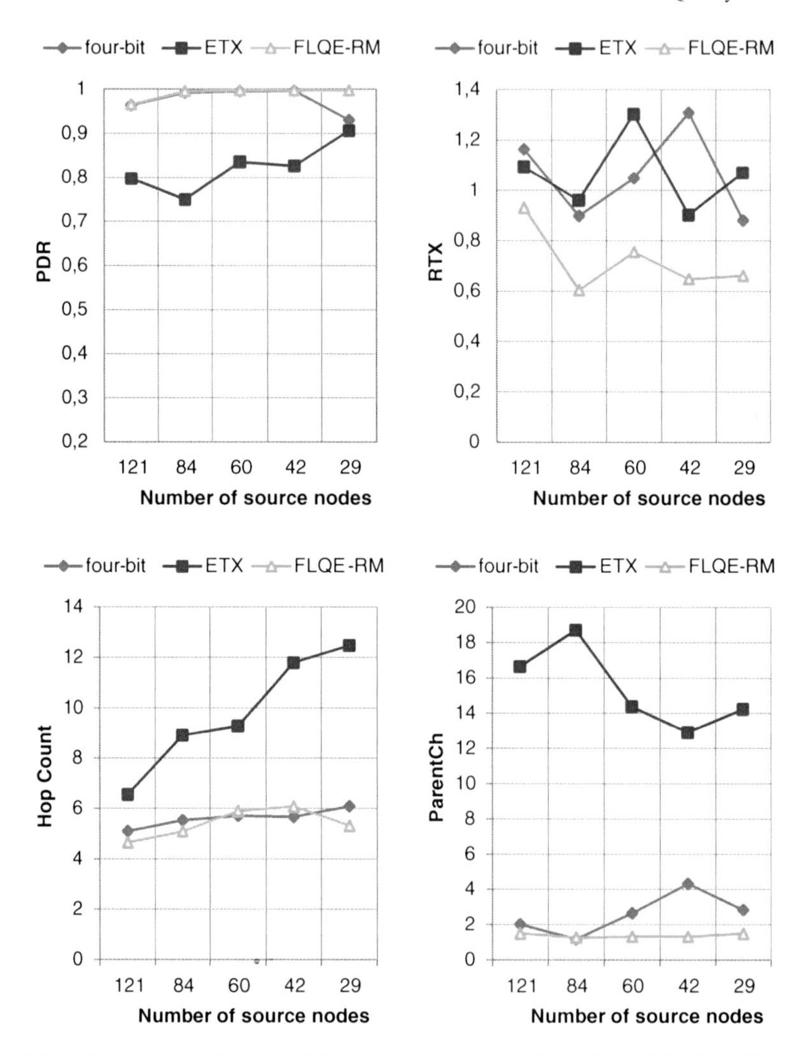

Fig. 5.4 Performance as a function of the number of source nodes (refer to Table 5.1—Set 3)

(precisely, TelosB mote) running FLQE-RM as routing metric consumes a total ROM footprint equal to 27.10 KB and a total RAM footprint equal to 4.47 KB. Compared to four-bit and ETX, FLQE-RM has more memory footprint as depicted in Table 5.5. Nevertheless, today's sensor platforms provide higher memory than that consumed by FLQE-RM. For example, a TelosB mote has total ROM of 48 KB and a total RAM of 10 KB. Our experimental study with Indriya and MoteLab proves that the FLQE-RM metric can be implemented on TelosB and TMote Sky motes.

FLQE-RM relies on F-LQE estimator, which is computationally more complex than four-bit and ETX. Typically, F-LQE computes four link quality metrics (SPRR, ASL, SF, and ALQI) applies these metrics to piecewise linear membership functions,

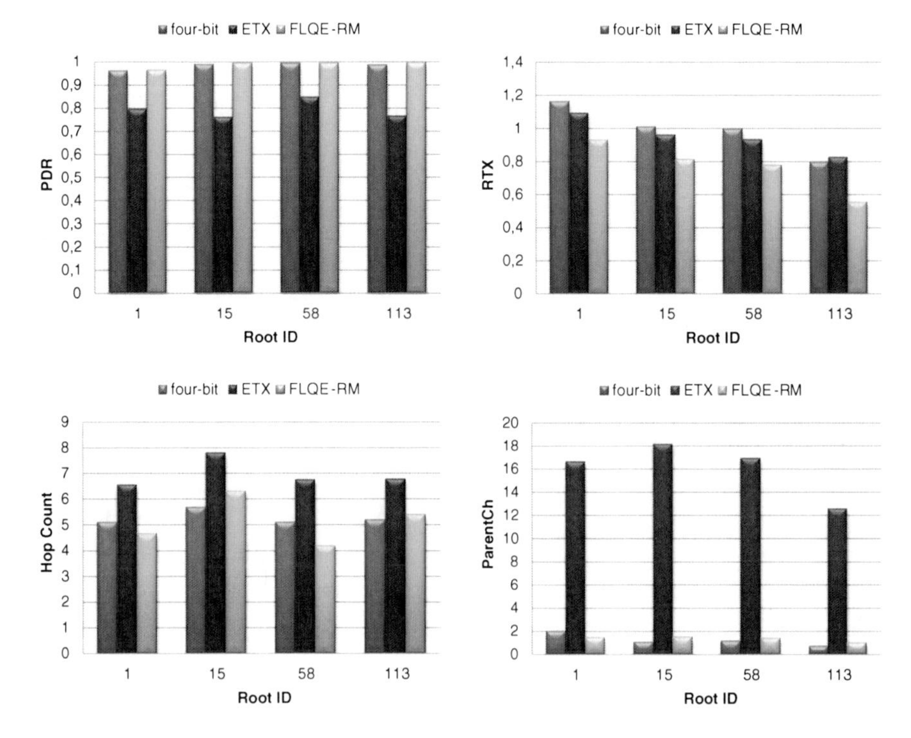

Fig. 5.5 Performance as a function of the topology (refer to Table 5.1—Set 4)

Table 5.5 Memory footprint of four-bit, ETX, and FLQE-RM

	Four-bit	ETX	FLQE-RM
ROM (in KB)	22.28	22	27.10
RAM (in KB)	4.04	4.06	4.47

then combines the different membership levels into a particular equation. On the other hand four-bit combines two link quality metrics through a simple weighted sum (the EWMA filter), and ETX uses a single link quality metric.

5.2.5 Discussion

FLQE-RM, four-bit, and ETX build routing trees, based on link quality estimation. Typically, an efficient routing metric (i) reduces the number of packet transmissions and retransmissions in the network, (ii) increases its delivery and (iii) ensures a stable topology. Our experimental study demonstrates that FLQE-RM establishes and maintains the routing tree better than four-bit, and ETX as it generally presents the highest PDR and the lowest RTX, Hop count and ParentCh. The effectiveness of FLQE-RM as a routing metric can be interpreted by (i) the accuracy of the link quality estimation as well as (ii) the efficiency of path cost evaluation.

In the context of CTP routing, all routing decisions are based on link quality estimation. Therefore, the accuracy of link quality estimation significantly impacts the effectiveness of routing metrics: the more accurate the estimate is, the better routing decisions are. In the previous chapter, we have shown that F-LQE is more accurate than four-bit and ETX as it provides a fine grained classification of links, especially intermediate links (these are the most difficult to assess). Thus, our experimental results confirm the accuracy of F-LQE, which is traduced by the correctness of routing decisions.

The effectiveness of a routing metric depends not only on the accuracy of link quality estimation, but also on how to use link estimates to evaluate the path cost. The FLQE-RM path cost function allows to select paths constituted with high quality links, while avoiding those having some weak links among high quality links. This path cost function also favor the selection of short paths. In fact, the path cost functions of four-bit and ETX metrics also shares these features. That is, they take into account the path global quality and implicitly favors the selection of short paths that do not have poor links. Hence, what makes FLQE-RM more effective than four-bit and ETX is the accuracy of link quality estimation through the use of F-LQE.

Our experimental results also shows that four-bit performs better than ETX. This result mainly pertains to the accuracy of link quality estimation. Four-bit takes into account more link aspects compared to ETX, as it combines RNP and estETX. estETX is a smoothed ETX using the EWMA filter.

The better performance of FLQE-RM over four-bit and ETX does not come without a price. As we have shown above, FLQE-RM involves higher memory footprint and computation complexity.

5.3 On the Use of Link Quality Estimation for Mobility Management

5.3.1 Link Quality Estimation for Mobile Applications

Nowadays, mobility is one of the major requirements in several emerging ubiquitous and pervasive sensor network applications, including health-care monitoring, intelligent transportation systems and industrial automation [15–17]. In some of these scenarios, mobile nodes are required to transmit data to a fixed-node infrastructure in a timely and reliable fashion. For example, in clinical health monitoring [18, 19], patients have small sensing devices embedded in their bodies that report data through a fixed wireless network infrastructure. In these type of scenarios, it is necessary to provide a reliable and constant stream of information.

Mobility management is a wide area which covers various aspects such as handoff process, re-routing, re-addressing and security issues. Due to the scope of this book, our main focus is on the *handoff process*. Handoff refers to the process where a mobile node disconnects from one point of attachment and connects to another. Hence, it is clear that handoff process greatly relies on link quality estimation. This fact motivates us to address the question of how to use link quality estimation for an efficient handoff in mobility management solutions.

In mobile applications, especially those deployed in harsh environments with rapid variations of wireless channel, what is important for a fast handoff decision is not just the accuracy of link quality estimation, but also the possibility to gather instantaneous link quality estimation at the time of transmission. However, accuracy and responsiveness in link estimation are two conflicting requirements. As discussed in Chap. 3, accurate link quality estimation requires the combination of several link properties to provide a snapshot of the real link status. Such combination of several link quality metrics into one composite LQE is time consuming, because it requires averaging over several link measurements (e.g. to compute PRR). Consequently, a composite LQE may not able to provide timely link quality estimation, which has a negative impact on the effectiveness of the handoff schema.

We argue that in mobile applications, single-metric LQEs such as RSSI and SNR, which are considered not sufficiently accurate [6], have the advantage of being responsive and thus would be more appropriate for handoff process. However what LQE to use? and how to tune it for a fast handoff? are just inevitable questions that are addressed next.

5.3.2 *Overview of Handoff Process*

A naive handoff solution in applications with mobile users is to broadcast information to all neighboring static nodes, known as access points (APs), within the transmission range. Broadcasts lead to redundant information at neighboring APs. This also implies that the fixed infrastructure wastes resources in forwarding the same information to the end point.

A more efficient solution for mobile nodes is to use a single AP to transmit data at any given time. This alternative requires nodes to perform reliable and fast handoffs between neighboring APs. In practice, a handoff starts when the link with the current (serving) AP drops below a given value (TH_{low}) and stops when it finds a new AP with the required link quality (TH_{high}). The most important issues that should be considered when designing a handoff mechanism for low-power networks are as follows:

5.3.2.1 Types of Handoff

The type of handoff is dictated by the capabilities of the radio, standards and technologies. Handoffs are classified into two main categories: hard handoffs and soft handoffs.

The *soft handoff* technique in wireless cellular networks uses multiple channels at the same time. This characteristic enables a mobile node to communicate with several APs and assess their link qualities while transmitting data to the serving AP. It is possible to perform soft handoff by utilizing a network-based mobility management which is supported by mobile IPv6. The use of IPv6 imposes extra overhead and increases the energy consumption of the network drastically. The implementation of soft handoff approach is feasible for low-power wireless networks, however, it is impractical for many applications.

In a *hard handoff*, the radio can use only one channel at any given time, and hence, it needs to stop the data transmission before the handoff process starts. Consequently, in hard handoffs it is central to minimize the time spent looking for a new AP. Low-power nodes typically rely on low-power radio transceivers that can operate on a single channel at a time, such as the widely used CC2420. This implies that current low-power wireless networks should utilize a hard handoff approach.

5.3.2.2 Impact of Low-Power Links on Handoff

Low-power links have two characteristics that affect the handoff process: short coverage and high variability [20].

Short coverages imply low densities of access points. In cellular networks, for example, it is common to be within the range of tens of APs. This permits the node to be conservative with thresholds and to select links with very high reliability. On the other hand, sensor networks may not be deployed in such high densities, and hence, the handoff should relax its link quality requirements. In practice, this implies that the handoff parameters should be more carefully calibrated within the (unreliable) transitional region.

The high variability of links has an impact in stability. When not designed properly, handoff mechanisms may degrade the network performance due to the ping-pong effect, which consists in mobile nodes having consecutive and redundant handoffs between two APs due to sudden fluctuation of their link qualities. This happens usually when a mobile node moves in the frontiers of two APs. Hence, to be stable, a handoff mechanism should calibrate the appropriate thresholds according to the particular variance of its wireless links.

5.3.2.3 Handoff Triggering

The first step in a handoff scheme is to determine when should a node deem a link as weak and start looking for another AP. We call this step, which is totally based

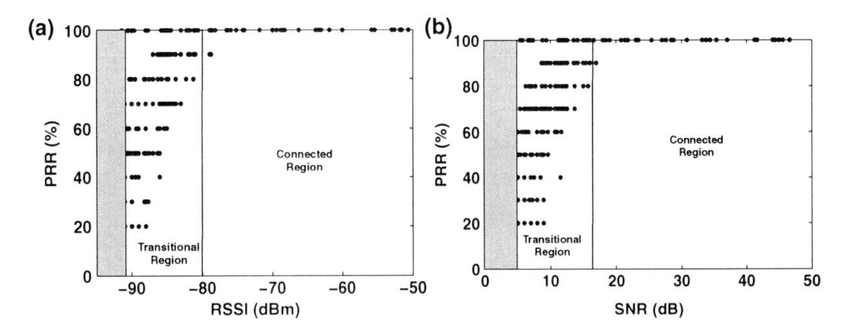

Fig. 5.6 Low-power link model. **a** RSSI versus PRR. **b** SNR versus PRR

on link quality estimation, handoff triggering. In the sensor network community, the *de-facto* way to classify links is to use the connected, transitional and disconnected regions (refer to Chap. 1 for the description of these regions).

We use RSSI and SNR in order to identify these regions. Hence, we gathered RSSI and SNR values at different parts of a building utilizing different nodes. The results illustrated in Fig. 5.6 have been collected by sampling many signals (every 10 ms) during a mobile node (MN) trip from one AP to another AP (with transmission power of −25 dBm). The figures depict these three regions for RSSI and SNR [19] that agree with studies in [21]. The SNR is calculated by measuring the noise-floor immediately after receiving the packet, and then, subtracting it from the RSSI value. The RSSI regions can be mapped directly to the SNR ones by subtracting the average noise-floor. The graphics illustrate that in a transitional region, the RSSI values are in a range of [−92 dBm, −80 dBm] and the SNR values are in a range of [5 dB, 17 dB].

5.3.2.4 Handoff Parameters

The process of switching from one AP to another should be performed wisely such that the ping-pong effect is minimized. Figure 5.7 depicts the two cases of efficient and inefficient handoff mechanism. In this example, the experiment encompasses two APs and a mobile node. The y-axis shows the RSSI detected by the serving AP and the vertical bars denote the handoffs performed. Note that, the RSSI is measured at the AP side. The transitional region in sensor networks, for the CC2420 radio transceiver, encompasses the approximate range (shown in Fig. 5.6). Intuition may dictate that it is better to perform handoff in the connected region with more reliable links. A conservative approach is depicted in Fig. 5.7a, which considers −85 dBm as the lower threshold (TH_{low}), and the upper threshold (TH_{high}) is 1 dB higher. These parameters lead to a negative effect: a long delay that takes three handoffs for a mobile node moving between the two contiguous APs (ping-pong effect). Figure 5.7b shows that by considering a wider margin, deeper into the transitional region, the ping-pong effect disappears and the delay is reduced a lot. This mechanism which

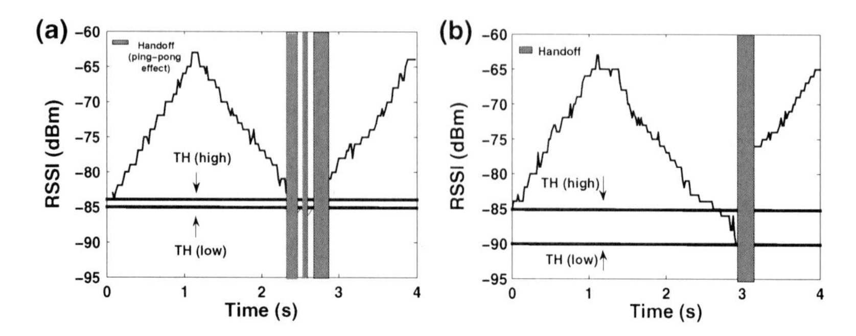

Fig. 5.7 **a** An example of an inefficient handoff. **b** An example of an efficient handoff [19]

involves a disconnection period is an example of the case where the MN has one radio and does not support IP. A careful calibration of the parameters can reduce the disconnection period which is so called the handoff delay.

There are various parameters involved in a handoff process which are supposed to be defined on the MN and AP devices. The threshold levels and the hysteresis margin are the most important parameters which define the starting and ending moments of a disconnection. The lowest threshold has to consider the boundaries of the transitional region. If the threshold is too high, the node could perform unnecessary handoffs and if the threshold is too low the node may use unreliable links. If the margin is too narrow, the mobile node may end up performing unnecessary and frequent handoffs between two APs (ping-pong effect). If the margin is too wide, the handoff may take too long which ends up increasing the delay and decreasing the delivery rate.

5.3.3 Soft Handoff in Low-Power Wireless Networks

As previously described, there are two major strategies to make a handoff process that are soft handoff with network layer solution and hard handoff with MAC layer solution. The first approach neglects the energy efficiency issue has been extended in [22, 23].

In [22] the problem related to the mobility of sensor node (SN) to handoff between different gateways (GW), connected to the backbone network is addressed. It proposes a soft handoff decision for low-power wireless networks based on 6LoWPAN (SH-WSN6) which avoids unnecessary handoffs when there are multiple GWs in the range of SNs. The sensor node is able to register to multiple GWs at the same time by using Internet Protocol (IP) solution. The SH-WSN6 takes advantage of router advertisement (RA) message defined in the Internet control message protocol (ICMP). GWs transmit RA messages periodically to advertise their presence. At first, SN can register to only one GW. By receiving RA in each interval, the SN decides for the best GW. Every time a SN registers with a new GW, it gains a new route.

This improves connectivity by having route diversity. If there is an unreliable link, comparison algorithm makes a decision to remove that link and therefore improves the QoS since poor links will not be used anymore. Comparison algorithm makes independent decision for start of handoff. Decision is made based on the comparison of the ratio of RA messages coming from GWs in the range. SN also notices when a GW moves away from SN's range by comparing the ratio of RA messages. Comparison algorithm assumes that GW's send RA messages at the same rate, which is a reasonable assumption.

GINSENG project presents a hard handoff solution within its mobility operation [23]. It is implemented on top of GinMAC that is a TDMA scheme for channel access with a pre-dimensioned virtual tree topology and hierarchical addresses. Two control messages are transmitted in order to support the attachment of the MN to a new point of attachment. These messages are the Join and the Join Ack that are sent/received when the MN is still attached to the previous tree position. Therefore, the role of the dynamic topology control in soft handoff mobility is to support the re-attachment of the MN to a different tree position as a result of movement inside the testbed area. In the handoff decision rules, some parameters are defined that are (i) RSSI threshold, (ii) better RSSI, (iii) number of lost packets, and (iv) packet loss percentage. These values are set according to the application requirements.

5.3.4 Hard Handoff in Low-Power Wireless Networks

The second approach in doing a handoff addresses a MAC layer solution for hard handoff mechanism in mobile low-power wireless networks. This solution is either specialized for passive decision with non-real-time support in [18] or for active decision with real-time support in [19].

In [18] authors describe a wireless clinical monitoring system collecting the vital signs of patients. In this study, the mobile node connects to a fixed AP by listening to beacons periodically broadcasted by all APs. The node connects to the AP with the highest RSSI. The scheme is simple and reliable for low traffic data rates. However, there is a high utilization of bandwidth due to periodic broadcasts and handoffs are passively performed whenever the mobile node cannot deliver data packets.

Smart-HOP [19] is a fast handoff process for low-power wireless networks, which get advantage of high responsiveness of RSSI/SNR and embeds a software based approach to reduce the decision inaccuracy. This is provided by adding three features: (i) getting averaged value of RSSI/SNR by exploiting a sliding window to minimize the sudden changes, (ii) filtering out the asymmetric property by using reply packets at the Data Transmission and Discovery Phase, and (iii) applying wide hysteresis margin to reduce the link variability in the transitional region.

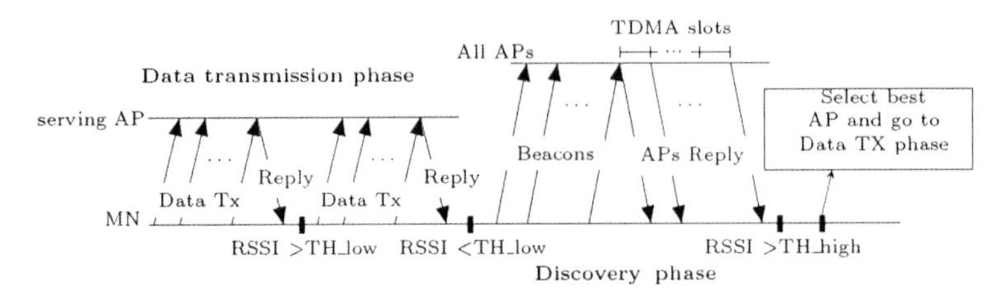

Fig. 5.8 Time diagram of the smart-HOP mechanism [19]

5.3.5 Smart-HOP Design

The smart-HOP algorithm has two main phases: (i) *Data Transmission Phase* and (ii) *Discovery Phase*. A timeline of the algorithm is depicted in Fig. 5.8.

Initially, the mobile node is not attached to any access point. This state is similar to the case when the MN disconnects from one AP and searches for a better AP. In both cases, the MN performs a Discovery Phase by sending n request packets in a given window w and receiving a $reply$ packet from each neighboring AP. The reply packet embeds the link quality level that is defined as the average RSSI/SNR level of n consecutive packets. By receiving reply packets at the MN, the down-link information is extracted. Then the mobile nodes selects the AP with highest link quality level which in turn considers the asymmetry feature of low-power wireless links. Upon detecting a good link, the MN resumes a Data Transmission Phase with the AP serving the most reliable link. The data packets are sent in burst and receive a reply afterwards similar to the Discovery Phase. This process enables monitoring the current link during the normal data communication process. The details of both phases are shown in Fig. 5.8. The smart-HOP process relies on three main tuning parameters, which are presented in details as follows.

Parameter 1: link monitoring frequency. It is an important parameter for any handoff process, which determines how frequent the link monitoring should be. The link monitoring property is captured by the window size parameter (ws), which represents the number of packets required to estimate the link quality over a specific time. A small ws (high sampling frequency) provides detailed information about the link but increases the processing of reply packets, which leads to higher energy consumption and lower delivery rates. The packet delivery reduces as the MN opts for several unnecessary handoffs. The handoff is ordered by detecting low quality links that happen by sudden fluctuations of signal strength. On the other hand, a large ws (low sampling frequency) provides only coarse grained information about the link and decreases the responsiveness of the system. A large ws leads to late decision, which is not suitable for a mobile network with dynamic link changes.

The mobile node starts the Discovery Phase when the link quality goes below a certain threshold (TH_{low}) and looks for APs that are above a reliable threshold ($TH_{high} = TH_{low} + HM$, where HM is the hysteresis margin). During the Discovery Phase, the mobile node sends ws beacons periodically and the neighboring

APs reply with the average RSSI or SNR of the beacons. If one or more APs are above TH_{high}, the mobile node connects to the AP with the highest link quality and resumes data communication, else, it continues broadcasting beacons in burst until discovering a suitable AP. In order to reduce the effects of collisions, the APs use a simple TDMA MAC.

Parameter 2: threshold levels and hysteresis margin. In low-power wireless networks, the selection of thresholds and hysteresis margins is dictated by the characteristics of the transitional region and the variability of the wireless link. The lowest threshold has to consider the boundaries of the transitional region. Wireless sensors spend most of the time in the transitional region. The exact threshold level within the transitional region is computed from the simulation and experimental analysis. If threshold TH_{low} is too high, the node could perform unnecessary handoffs (by being too selective). If the threshold is too low, the node may use unreliable links. The hysteresis margin plays a central role in coping with the variability of low-power wireless links. If the hysteresis margin is too narrow, the mobile node may end up performing unnecessary and frequent handoffs between two APs (ping-pong effect), as illustrated in Fig. 5.7. If the hysteresis margin is too large, the handoff may take too long, which ends up increasing the network inaccessibility time, and thus delivery delay and decreasing the delivery rate.

Parameter 3: AP stability monitoring. Due to the high variability of wireless links, the mobile node may detect an AP that is momentarily above TH_{high}, but the link quality may decrease shortly after being selected. In order to avoid this, it is important to assess the stability of the AP candidate. After detecting an AP above TH_{high}, smart-HOP sends m further bursts of beacons to validate the stability of that AP. The burst of beacons stands for the ws request beacons followed by the reply packets received from neighboring APs. Stability monitoring is tightly coupled to the hysteresis margin. A wide hysteresis margin requires a lower m, and vice versa.

5.3.6 Smart-HOP Observations

To evaluate smart-HOP functionality, different scenarios were considered, which are summarized in Table 5.6. For example, scenario A with a 5 dBm margin and stability 2, means that after the mobile node detects an AP above $TH_{high} = -90$ dBm, the node will send two 3-beacon bursts to observe if the link remains above TH_{high}. The hysteresis margin HM captures the sensitivity to ping-pong effects, and the number of bursts m, the stability of the AP candidate (recall that each burst in m contains three beacons).

Calibrating the parameters of smart-HOP requires a testbed that provides a significant degree of repeatability. A fair comparison of different parameters is only possible if all of them observe similar channel conditions. In order to achieve this, a model-train in a large room is employed. The room is 7×7 m and the locomotive follows a 3.5×3.5 m square layout. The speed of the locomotive is approximately

Table 5.6 Description of second set of scenarios

Scenarios	TH_{low} (dBm)	HM (dBm)	m
A	−95	1, 5	1, 2, 3
B	−90	1, 5	1, 2, 3
C	−85	1, 5	1, 2, 3
D	−80	1, 5	1, 2, 3

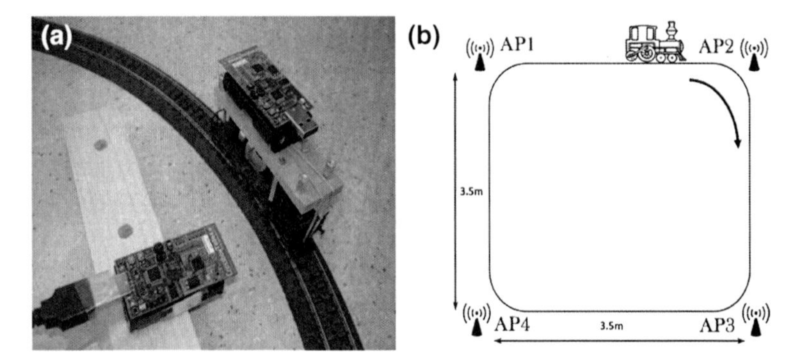

Fig. 5.9 **a** MN passing by an AP. **b** Nodes' deployment

1 m/s (average walking speed). Fig. 5.9a depicts a locomotive passing by an AP and Fig. 5.9b shows the experimental scenario.

In real-world applications, the deployment of access points (or base stations) is subject to an accurate study to ensure the coverage of the area of interest. In cellular networks, the density of access points guarantees full coverage and redundancy. In other wireless networks, the density of access points depends on the real-time requirements of the application. In critical applications, complete coverage is an essential requirement. To prevent extreme deployment conditions such as very high or very low density of APs, smart-HOP tests provided minimal overlap between contiguous APs. For each evaluation tuple $< TH_{low}, HM, m >$, the mobile node took four laps, which lead to a minimum of 16 handoffs. The experiments show the results for the narrow margin (1 dBm), and for the wide margin (5 dBm).

Figure 5.10 shows the number of handoffs, handoff delay and the relative packet delivery ratio for two cases of narrow and wide hysteresis margin.

The high variability of low-power links can cause severe ping-pong effects. Figure 5.10a, b show two important trends with narrow margin; first, all scenarios have ping-pong effects. The optimal number of handoffs is 16, but all scenarios have between 32 and 48. Due to the link variability, the transition between neighboring APs requires between 2 and 3 handoffs. Second, a longer monitoring of stability m helps alleviating ping-pong effects. Moreover, for all scenarios, the higher the stability, the lower the number of handoffs.

Thresholds at the higher end of the transitional region lead to longer delays and lower delivery rates. A threshold selected at the higher end of the transitional region can lead to an order of magnitude more delay than a threshold at the lower

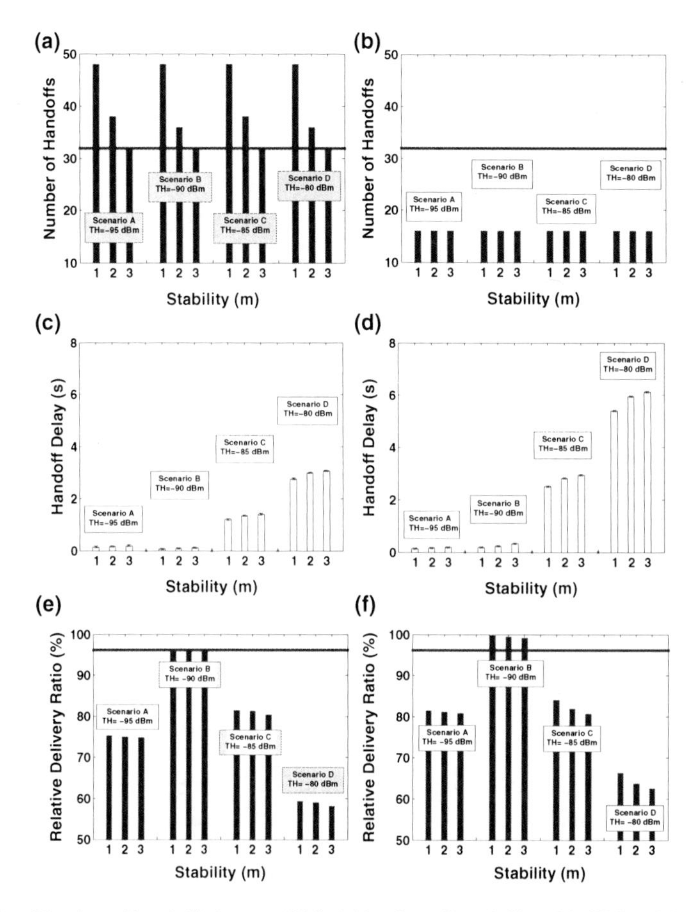

Fig. 5.10 **a** Number of handoffs (narrow HM). **b** Number of handoffs (wide HM). **c** Mean handoff delay (narrow HM). **d** Mean handoff delay (wide HM). **e** Relative delivery ratio (narrow HM). **f** Relative delivery ratio (wide HM). The *horizontal lines* represent the results for the best scenario: 32 for the number of handoffs and 96 for the relative delivery ratio [19]

end. This happens because mobile nodes with higher thresholds spend more time looking for overly reliable links, and consequently less time transmitting data.

The most efficient handoffs seem to occur for thresholds at the lower end of the transitional region with wide hysteresis margin. Scenario B (−90 dBm) with stability 1 maximizes the three metrics of interest. It leads to the least number of handoffs, with the lowest average delay and highest delivery rate. It is important to highlight the trends achieved by the wider hysteresis margin. First, the ping-pong effect is eliminated in all scenarios. Second, contrarily to the narrower hysteresis margin, monitoring the stability of the new AP for longer periods ($m = 2$ or 3) does not provide any further gains, because the wider margin copes with most of the link variability.

Impact of interference. The functionality of smart-HOP was also analyzed under interference, by comparing the RSSI and SNR based models. Different types of interference such as periodic (similar to microwave ovens) and bursty (similar to WiFi devices) were generated. The observations indicated that smart-HOP with SNR increases both the average delay and the delivery ratio. The longer handoff delay occurs because the MN spends more time in the Discovery Phase looking for good links. The MN detects the presence of interference earlier and starts the Discovery Phase. It attaches to the new AP by observing a high quality link in terms of lower noise-floor. Hence, in SNR based handoff with the necessity of always connecting to lower noisy link, the packet delivery rate is higher.

5.3.7 Conclusion

Link quality estimation is a fundamental building block for several network protocols and mechanisms, especially for routing and mobility management. The first part of this chapter addressed the problem of using link quality estimation for improving routing performance, especially CTP routing protocol. We have presented FLQE-RM, a routing metric based on F-LQE. Based on TOSSIM 2 simulation and real experimentation, FLQE-RM was found to improve CTP routing performance. Typically, FLQE-RM establishes and maintains CTP routing trees better than four-bit, and ETX.

The second part of this chapter addressed the problem of using link quality estimation for a fast handoff process in mobility management. Due to the high unreliability and dynamic changes of low-power lossy links with mobility support, a fast/responsive LQE is more acceptable than an accurate yet less responsive LQE. Smart-HOP proposes a hard handoff process for mobile low-power wireless network applications. It takes advantage of a sliding window to reduce the sudden fluctuations, filters out the asymmetry property of the link and applies a wide hysteresis margin to reduce the link variability. The results indicated that smart-HOP is able to perform a fast handoff with high delivery ratio.

References

1. Puccinelli D, Haenggi M (2010) Reliable data delivery in large-scale low-power sensor networks. ACM Trans Sens Netw 6(4):1–41
2. Woo A, Tong T, Culler D (2003) Taming the underlying challenges of reliable multihop routing in sensor networks. In: Proceedings of the 1st international conference on embedded networked sensor systems (SenSys '03). ACM, New York, pp 14–27
3. Couto DSJD, Aguayo D, Bicket J, Morris R (2003) A high-throughput path metric for multi-hop wireless routing. In: Proceedings of the 9th annual international conference on mobile computing and networking (MobiCom '03). ACM, New York, pp 134–146

4. Fonseca R, Gnawali O, Jamieson K, Levis P (2007) Four bit wireless link estimation. In: Proceedings of the 6th international workshop on hot topics in networks (HotNets VI), ACM SIGCOMM
5. Chen Br, Muniswamy-Reddy KK, Welsh M (2006) Ad-hoc multicast routing on resource-limited sensor nodes. In: Proceedings of the 2nd international workshop on Multi-hop ad hoc networks: from theory to reality, REALMAN '06. ACM, New York, pp 87–94
6. Baccour N, Koubâa A, Mottola L, Zuniga MA, Youssef H, Boano CA, Alves M (2012) Radio link quality estimation in wireless sensor networks: a survey. ACM Trans Sens Netw 8(4):1–34
7. CTP: Collection tree protocol. http://www.tinyos.net/tinyos-2.x/doc/html/tep123.html
8. LEEP protocol (2007) http://www.tinyos.net/tinyos-2.x/doc/html/tep124.html
9. Werner-Allen G, Swieskowski P, Welsh M (2005) Motelab: a wireless sensor network testbed. In: Proceedings of the international symposium on information processing in sensor networks (IPSN '05). IEEE Press, New York, pp 483–488
10. Doddavenkatappa M, Choon MC, Ananda AL (2011) Indriya: a low-cost, 3d wireless sensor network testbed. In: Proceedings of the 8th international ICST conference on testbeds and research infrastructures for the development of networks and communities
11. Handziski V, Köpke A, Willig A, Wolisz A (2006) Twist: a scalable and reconfigurable testbed for wireless indoor experiments with sensor networks. In: Proceedings of the 2nd international workshop on multi-hop Ad hoc networks: from theory to reality (REALMAN '06). ACM, New York, pp 63–70
12. Arora A, Ertin E, Ramnath R, Nesterenko M, Leal W (2006) Kansei: a high-fidelity sensing testbed. IEEE Internet Comput 10(2):35–47. doi:10.1109/MIC.2006.37
13. Johnson D, Flickinger D, Stack T, Ricci R, Stoller L, Fish R, Webb K, Minor M, Lepreau J (2005) Emulab's wireless sensor net testbed: true mobility, location precision, and remote access. In: Proceedings of the 3rd international conference on embedded networked sensor systems (SenSys '05). ACM, New York, pp 306–306. doi:10.1145/1098918.1098971
14. Puccinelli D, Gnawali O, Yoon S, Santini S, Colesanti U, Giordano S, Guibas L (2011) The impact of network topology on collection performance. In: Proceedings of the 8th European conference on wireless sensor networks, EWSN'11. Springer, Berlin, pp 17–32
15. Alemdar H, Ersoy C (2010) Wireless sensor networks for healthcare: a survey. Comput Netw 54:2688–2710
16. Qin H, Li Z, Wang Y, Lu X, Zhang W, Wang G (2010) An integrated network of roadside sensors and vehicles for driving safety: concept, design and experiments. In: Proceedings of IEEE PERCOM
17. Villaverde BC, Rea S, Pesch D (2012) InRout: a QoS aware route selection algorithm for industrial wireless sensor networks. Ad Hoc Netw 10(3):458–478
18. Chipara O, Lu C, Bailey TC, Roman GC (2010) Reliable clinical monitoring using wireless sensor networks: experiences in a step-down hospital unit. In: Proceedings of the 8th ACM international conference on embedded networked sensor systems (SenSys), pp 155–168
19. Fotouhi H, Zúñiga MA, Alves M, Koubâa A, Marrón PJ (2012) Smart-HOP: a reliable handoff mechanism for mobile wireless sensor networks. In: Proceedings of the 9th European conference on wireless sensor networks (EWSN), pp 131–146
20. Zúñiga MA, Irzynska I, Hauer JH, Voigt T, Boano CA, Römer K (2011) Link quality ranking: Getting the best out of unreliable links. In: Proceedings of the 7th IEEE international conference on distributed computing in sensor systems (DCOSS), pp 1–8
21. Srinivasan K, Kazandjieva MA, Agarwal S, Levis P (2008) The β-factor: measuring wireless link burstiness. In: Proceedings of the 6th international conference on embedded network sensor systems (SenSys '08). ACM, New York, pp 29–42
22. Petäjäjärvi J, Karvonen H (2011) Soft handover method for mobile wireless sensor networks based on 6LoWPAN. In: Proceedings of the 7th IEEE international conference on distributed computing in sensor systems (DCOSS)
23. Zinonos Z, Vassiliou V (2011) S-GinMob: soft-handoff solution for mobile users in industrial environments. In: Proceedings of the 7th IEEE international conference on distributed computing in sensor systems (DCOSS)

Chapter 6
Conclusions

Low-power wireless technologies have been major enablers for pervasive computing applications as diverse as environmental monitoring, smart buildings, plant process control, structural health monitoring and more generically any application under the smart cooperating objects and "Internet of Things" umbrella. However, the scalability requirements typically imposed by such applications constrain the cost/energy per node to unprecedented levels. Namely, system designers are usually limited to use low-cost radio transceivers operating at the unlicensed ISM spectrum and with transmission powers two orders of magnitude smaller than commodity wireless technologies, as well as light ultra-compacted application, data processing and middleware software and protocol stacks to cope with very limited memory and processing capabilities. Therefore, communications typically tend to be unreliable and, worse than that, unpredictable.

On the contrary, the quality-of-service requirements of some of these applications are demanding and are expected to be fulfilled. The need for multi-hop routing (particularly in large-scale systems), the harsh environmental characteristics and the application dynamics (e.g. imposed by mobility) in many of these application scenarios turn this problem even more acute. In this line, effects such as noise, interference, shadowing and multi-path distortion need to be investigated, characterized and, if possible, mitigated.

This book aims at providing an overall picture of radio link quality estimation and interference aspects in low-power wireless networks. In this context, we have identified the main characteristics of low-power links, mostly drawn from empirical observations. Then, we analyzed the coexistence of different wireless technologies and interference sources operating in the ISM band and particularly focused on interference measurement, modeling and mitigation techniques. The fundamental concepts and a taxonomy of link quality estimation have been presented, as well as some guidelines for system designers. Building upon the RadiaLE framework, we have made a comparative performance evaluation of the most popular link quality estimators, both based on simulation (TOSSIM) and experimental models. Finally, we elaborated on how link quality estimation can be efficiently tuned and integrated in higher level mechanisms, namely routing and mobility, such that their performance is improved.

N. Baccour et al., *Radio Link Quality Estimation in Low-Power Wireless Networks*, SpringerBriefs in Electrical and Computer Engineering,
DOI: 10.1007/978-3-319-00774-8_6, © The Author(s) 2013